The Patrick Moore Practical Astronomy Series

For further volumes:
http://www.springer.com/series/3192

Twenty-Five Astronomical Observations That Changed the World

And How To Make Them Yourself

Michael Marett-Crosby

 Springer

Michael Marett-Crosby
Mon Plaisir
Route de la Haule
Jersey JE3 8BD
United Kingdom

ISSN 1431-9756
ISBN 978-1-4614-6799-1   ISBN 978-1-4614-6800-4 (eBook)
DOI 10.1007/978-1-4614-6800-4
Springer New York Heidelberg Dordrecht London

Library of Congress Control Number: 2013939733

Cover illustration: Courtesy of NASA/JPL/Space Science Institute

Printed on acid-free paper

Springer is part of Springer Science+Business Media (www.springer.com)

Preface

This book is about bringing the far-off close. It offers twenty-five observations that can be achieved using modest amateur equipment. Each of these observations has changed how we understand our world and its place in the universe. Some have made headlines, transforming how we think about ourselves. Others have caused quieter revolutions. But each of these observations has a place in history.

And you can see them for yourself. With the help of this book, you can follow in the footsteps of the great astronomers and connect with the science, the history and the changes that have flowed from them. You can share with others why astronomy matters for us all.

Each chapter contains observations that require fairly simple astronomical tools: binoculars, a telescope, a sense of where things are in the sky and lots of patience! Each observation is made from familiar starting points, so there is no need to have go-to software for your telescope.

We will cover the science that explains and illuminates the observations so as to understand what we are seeing. We will also witness the impact that each observation has made on human history and culture, and see why a planet or tiny spot in the sky has affected the world in which we live.

How to Use This Book

Readers who are beginning their astronomical explorations can follow this book in order. The early chapters contain straightforward observations and introduce key scientific tools that help to understand the planets and the stars. More experienced observers will be able to see familiar objects in a new way, as well as seeking out more challenging targets. Everyone will discover things that they can share with non-astronomers, so that we can talk about our passion in a way that builds connections.

The history of astronomy is enfolded within the observations. Great figures appear several times and sometimes in surprising ways, as we see for ourselves the observations that they made.

This is inevitably a personal selection. It is formed by my own observing experience in the northern part of the northern hemisphere and by my perception of great moments in astronomical history. I make some suggestions as to what might be the 26th observation in the last chapter of the book. I would welcome other ideas.

About the Pictures

Many images in this book are the work of amateur astronomers. The pictures taken by the Hubble Space Telescope and other instruments are magnificent, but they can also mislead. If we think we are going to see the Horsehead or Crab Nebula as Hubble does through our telescopes, we are going to be disappointed. But amateur astrophotographers can, with great patience and skill, achieve amazing images.

Without exception, those whom I approached gave permission for their pictures to appear in this book. They want their work to be enjoyed.

Acknowledgements

Many people have contributed to bringing together this book.

I would like first of all to thank John Watson and Maury Solomon at Springer, who between them helped me to hone a plan for this book that has remained robust and is, I hope, something like what they want.

It is a pleasure to thank all those who have contributed their images so willingly. Individuals are credited beneath their work. I would like here to acknowledge those who have provided multiple images and who gave me other assistance: John Ambrose, Dr Jean-Marc Bonnet-Bidaud, Paul Downing, Kathryn McKee, Dr Birgit Krummheuer, Jim Misti, Steven Ringwood, Professor Paul Schenk, George Tarsoudis, Christian Viladrich and Jimmy Westlake.

All maps have been created with the Cartes du Ciel software. The program makes use of the VizieR catalogue access tool, CDS, Strasbourg, France.

Throughout the writing of this book, I have been encouraged and inspired by members of the Highlands Astronomical Society, who have shared so willingly their interests and expertise: Arthur Milnes, Pat Williams, Paul Jenkins, Pauline Macrae, Pat Escott, Gordon McKenna, John Rosenfield for his infectious enthusiasm for Mercury, and Gerry Gaitens for the tutorial with the Lunt telescope.

I am particularly grateful to Eric Walker and Maarten de Vries for their help with images and to Antony McEwan for his advice on several of the observations.

My interest in astronomy began as a child when Adrian and Lucinda Phillips gave me my first telescope. I will always remain grateful for all that they have given me.

I would like to thank my family for all their support, in so many ways, during this last year and every year.

And also to thank James, without whom there is no way this book would even have started, let alone come to its end.

Contents

1 **The Eyes of the Moon** 1
   A Unique Sight 1
   First Observation 3
   Understanding What You've Seen 6
   Glimpsing Earthshine 7
   Meeting Copernicus 8
   The Moon Has a History 10
   The Moon Is Made of Alabaster? 10
   New Ways of Thinking 12

2 **How to See the Sun** 15
   How to Look at the Sun 16
   Looking at Sunlight 17
   Finding Helium 17
   The White Light of the Sun 18
   A Funeral for Pseudo-Philosophy 20
   The Boiling Sun 21
   The Wind in Our Hair 22
   Sound and Glory 25

3 **How Stars Work** 27
   Finding the Dipper 27
   The Science of a Single Star 29
   Horses and Riders 31
   On the Move 33
   The Rest of the Bear 35
   Why a Bear? 37
   A Muddle of Stars and People 37

4   The Telescope Revolution 39

The Jeweled Necklace 40

Paths Never Trod Before by the Human Mind 41

Moving Stars 44

Galileo Condemned 45

From Galileo to Galileo 46

The Fires of Io 47

Where There Is Water… 49

The Pleasure of Telescopes 50

5   The Hunter's Stars 53

Making Sense of the Stars 54

The Hunter 56

Orion's Prey 60

Baptizing the Stars 62

6   Home and Next Door 63

Walking the Milky Way 63

Observing Andromeda 68

A Battle for the Stars 70

How Big Is Home? 71

The Shapes of Galaxies 72

7   Our Red Neighbor 75

Does Mars Move Backwards? 76

Mars Through a Telescope 78

Seeing Red? 82

Areophobia 82

Mariners and Vikings 83

Following the Water 85

Is There Life on Mars? 87

8   Future Suns 89

Observing the Orion Nebula 89

A Chaos of Future Suns 94

New Stars? 97

From Dry Plates to Hubble 98

Beware of Pictures? 101

9   Bright Dog and Dark Companion 103

How to Find Sirius 103

What We Are Seeing 107

Such Tiny Little Things 109

The Dog-Day Star 112

The Size of a Nose 113

10 Looking for Footsteps 115
A Trilogy of Craters 115
The Science of Craters 118
Dry Seas 119
Apollo in the History of Humanity 122
Climbing Mountains 123
Where Does the Moon Come From? 125
What About My Writing 'A City of the Moon?' 126

11 A Whispering River 129
The Young River 129
Where Does Spock Live? 131
The Boat and Its Galaxies 131
The Mouth of the River 136
Filling the Void 137
No Oak Tree 138

12 The Lights of Kings 139
Observing Comets 139
Comet Rain 142
Close Encounters 143
What Is a Comet? 145
A Hint of Life 147
Proving the Invisible 147
The Man of Science 149
The Art of Comets 150

13 Wonderful Demons 153
Distinguishing Variables 154
Meet the Demon 154
Of Kings and Queens 158
Standard Candles 160
Whale Watching 162
The Wonderful Star 162
Night-Gaunts 165

14 The King of the Planets 167
Observing Jupiter 167
What Is a Gas Giant? 170
Probing Jupiter 172
How the Planets Move 173
Delayed Light 174
Big Planet or Small Star? 176
Facts and Poetry 177

15 Meet the Gangs.. 179
The Sisters.. 179
Between Two Donkeys 182
Great Globulars 183
Finding the Edge 185
The Seated Queen 186
A Double Swarm of Stars 187
Sharing the Stars 189

16 Life and Death... 191
Asteroids, Planets and Dwarfs 191
Hunting Asteroids 193
Rubble Science.. 196
Up Close.. 199
Even Closer 200
Impact 200
Don Quixote Rides Again? 203

17 Science, Fear and Supernovae 205
New and Super-New 206
Meet a Supernova.. 208
I Discovered a Nebula!.. 209
Lifting the Veil 211
Little Green Men... 212
Feared Guests.. 213

18 Habituated to the Vast 217
Many a Dish of Coffee.. 217
Observing Uranus 219
What Is Uranus Like? 220
The Satellites of Uranus... 223
Finding Neptune... 223
Bringing It Back Together.. 225
Fear and Wonder 226

19 Monsters of Magnitude and Invisible Nothings 229
Black Swan 229
No Escape 231
Among the Hunting Dogs 232
Black Holes and Galaxies 234
In These Our Sight Plunges 235

20 Falling in Love.. 239
Observing Saturn and Its Rings 239
Observing Saturn's Moons... 241
Probing Saturn's System 243
Saturn in the Spring 244

This Most Deceitful Star 245
Enceladus Lives 246
A Landing on Titan 248
Two Cultures or One? 250

21   **Into the Deep** 253
The Most Important Star in the Sky.. 253
Blue Spirals and Active Cores 254
Why Is the Night Sky Dark? 256
Galaxies and More Galaxies 257
The Matter of Darkness 260
Back to Virgo 261
Nothing But a Gnab Gib 263

22   **Nearest the Sun**... 265
Seeing Venus 265
Transiting the Sun 268
Getting Below the Clouds 270
Cloud-Life? 272
Observing Mercury 272
MESSENGER at Mercury 273
Party Lights 276

23   **Round and Round** 277
Stories and Scribbles... 277
Observing Satellites 279
The Asteroid That Wasn't 281
We Ain't Gonna Do It with the Tools We've Got 281
And Now? 283
A Little Hope 284

24   **Finding Planets Around Other Stars** 287
The Twin's Friend 288
Doppler Spectroscopy 289
Using the Square 290
Precise Measurement and Direct Sight 291
The Lights Go Down.. 293
How Many Exoplanets Are Out There?.................................... 294
Goldilocks... 294
The Kepler Mission 296
Is This the One? 298

25   **Watching the Sunrise**... 299
The Wild South 299
The End of the Old World.. 302
Luna Incognita 302

Newton's Puzzling Moon.. 304
The Far Side of the Moon 305
The Moon's Dominion.. 307
Darkness and Light 309

About the Author 311

Index.. 313

Chapter 1

The Eyes of the Moon

Observation: Tycho and Copernicus craters
Significance: The Moon has a history, the Copernican revolution
Science: Lunar orbit, phases, features

When your new telescope was unpacked and set up on its mount, what did you do? The answer to this question is the same for many – pointed it towards the Moon. Any telescope or pair of binoculars can do that awe-inspiring thing and take us, while we stand on Earth, to the surface of a different world.

The Moon is where our journey starts through these 25 observation that have changed human history. We will return to it in later chapters, where we will explore in detail its many types of craters, seas, mountains and ridges. We will look for evidence of water and, most transient of all, for human footprints. But in this first chapter we are going to do some more basic observations of the Moon, seeking to understand why it's there, why it matters and what forces account for the contrasting light and shade that forms the face looking back at us.

A Unique Sight

The Moon is so familiar that we forget it is exceptional, and that the Earth-Moon combination is something very special in our Solar System. As astronomers have learned more about the classical planets from Mercury to Neptune, about dwarf planets beyond and between them, about asteroids and comets and planets circling other stars, it has become increasingly clear that the Moon is an unusual companion.

The Moon is the only natural satellite of Earth. Since English was first spoken, "the Moon" has referred to that one companion in the sky. In the seventeenth

M. Marett-Crosby, *Twenty-Five Astronomical Observations That Changed the World: And How To Make Them Yourself*, The Patrick Moore Practical Astronomy Series, DOI 10.1007/978-1-4614-6800-4_1, © Springer Science+Business Media New York 2013

century, the meaning of the word was expanded to include the natural satellites of other planets, but the capital letter distinguishes Earth's Moon from the moons of other planets. The adjectives used for things of Earth's Moon are *lunar*, from the Latin word *luna*, or rarely *selenian* or *selenic*, derived from Greek.

All the planets save two – Mercury and Venus – have natural satellites, two in the case of Mars, at least 64 for Jupiter. As new satellites are identified by remote probes around the giant planets, the population of our Solar System increases, and the Earth-Moon mini system seems to become more special. Earth is a little less than four times the size of the Moon, a pairing of near equals.

The largest other satellite of a planet in our Solar System is the moon Ganymede, in orbit around Jupiter. Ganymede is significantly larger than our Moon and larger than the planet Mercury as well, but its diameter is miniscule compared to the planet it circles. A similarly extreme relationship exists between Mars and its tiny moons Phobos and Deimos.

The only comparable configuration to the Earth-Moon lies far out in the icy edges of our Solar System, in a region called the Kuiper Belt. There we find the dwarf planet Pluto, whose largest satellite Charon has a ratio to Pluto much like that of our Moon to Earth, about 1:4. But the similarities end there. Both Pluto and Charon are small, and the Pluto system contains at least three other bodies.

Fig. 1.1 This is who we are. The Earth and the Moon as seen by Messenger, 17 August 2010 (Credit: NASA/Johns Hopkins University Applied Physics Laboratory/Carnegie Institution of Washington)

At the moment, and until our studies of planets around other suns offer this kind of detail, Earth and the Moon will remain a unique pair.

This image makes the point. Taken in March 2011 by the Messenger spacecraft as it prepared to enter orbit around Mercury some 114,000,000 miles away, it shows Earth and the Moon as two merging spots of light separated from each other by a tiny strip of darkness.

First Observation

Just looking at the Moon on a reasonably clear night is where astronomy starts. It was the beginning of humanity's encounter with the skies. And it revealed some important facts about our satellite.

First of all, the Moon follows a predictable course from east to west, different every night, growing in apparent size from a thin sliver to full, when a whole disk is on show. Second, this disk contains a pattern of brighter and more shaded areas – these have for many centuries been identified as either a face or a full human figure. "I was the Man i' the Moon, when time was," Stephano says in Shakespeare's *The Tempest*.

This pattern on the lunar surface makes an important scientific point. It enables us to distinguish between the darker areas, lunar seas or maria, and the rest of the terrain.

Fig. 1.2 Full Moon (Credit: Jim Misti, http://www.mistisoftware.com/astronomy/)

We are going to focus now on the Tycho Crater, the bright white mark towards the bottom of the lunar disk as we look at it with the naked eye. It's probably easiest to explore first with binoculars because with them, the crater will appear in the same place as it does to the naked eye. This is not true when using telescopes, most of which invert images. For the majority of astronomical targets this does not matter, but on the Moon, our sense that north is at the top becomes confused.

Tycho is one of few features to survive the glare of a full Moon. The shape of the crater is visible and so, too, are the bright white lines emerging from it. But in these conditions there is too much light to reveal the detail. Tycho shows much more when it is emerging from the darkness on an eight- or nine-day Moon.

This is a key skill in lunar observing. The lunar day lasts about 29½ of our Earth days, and so the line between dawn and dusk moves quite slowly as we observe it. At this line, the terminator, shadows are lengthened by the angle of sunlight, making features stand out much more clearly. It's possible to watch the lunar morning in real time.

The lunar surface is a complicated place, but Tycho is a large crater some 86 km in diameter and nearly 5 km deep. Depending on what time of night you are viewing, you should be able to see the long shadow cast by the crater wall that faces the rising Sun. These ramparts are 4.8 km high, nearly six times higher than the tallest skyscraper on Earth. The shadows will slowly reveal Tycho's central peak standing 2.25 km

Fig. 1.3 (Credit: NASA Goddard/Arizona State University)

above the crater's floor. Walls and a central peak are characteristic of complex lunar craters – the peak is shown in exceptional detail in this image taken by NASA's Lunar Reconnaissance Orbiter on June 10, 2011.

In the area immediately around the crater, you should also be able to spot a dark band. This feature, called an annulus, circles the crater – its name is drawn from the Latin for *little ring*.

With the crater in the center of a binocular image, we can move from Tycho to its neighborhood, a broken landscape of shattered craters. Just beneath Tycho lies the crumbling crater Street, while to Tycho's east it is worth exploring Pictet through a telescope. It has a fine pattern of hills at its center. The rays of Tycho have fallen over Pictet, giving it a mottled appearance under the right light.

These rays are Tycho's most distinctive feature. They are lines of material thrown out when the crater was gouged out of the lunar surface. They act as useful pointers to other features. Beginning with Tycho's clear double ray, follow it to the Mare Nubium, the Sea of Clouds.

Continuing anticlockwise, the next ray points almost directly at the Moon's south pole, bisecting the crater Clavius, while the next and longest ray stretches across as far as the Mare Nectaris, the Sea of Nectar. This mare lies between two much larger seas that show up as dark patches under the naked eye: Fecunditatis (Fertility) and Tranquilitatis (Tranquility). It was on the southern 'shore' of Mare Tranquilitatis that Apollo 11 landed on July 21, 1969.

It is well worth repeating the Tycho observation on successive nights. It helps in acquiring a sense for lunar topography, and it is possible to watch the advancing terminator create new shadows across features. Find Tycho at the other end of the lunar cycle, the last quarter of the Moon, for as Tycho slides into the night, you can watch the dusk fall for yourself.

There are two items that really help with lunar observing, be it a straightforward or a tricky target that you are seeking. One is a lunar map. This is easy to obtain and makes navigating across the lunar surface much easier. Take one out with you, protected from the dew or ice in a sealed plastic wallet. There are excellent lunar maps available for handheld tablets, and these show where the terminator lies, making lunar navigation easier than it has ever been.

It's also worth investing in a lunar or planetary filter. The Moon shines very brightly, and a quick way of reducing glare is to screw an appropriate filter onto the eyepiece. The effect is to make details and lunar shadows come alive under an altogether more kindly light.

We cannot end our look at Tycho without acknowledging something will be a disappointment. However hard you look, you will not be able to observe AMT1, the black monolith that appears in the crater in the opening of Arthur C. Clarke's and Stanley Kubrick's film *2001: A Space Odyssey*. If Tycho seems familiar, that is where you have seen it before, for the film does a good job of making crater feel real. If it rains every time you try to look at the Moon, you can always watch the film instead.

Understanding What You've Seen

One of the central themes of this book is that understanding the science increases the pleasure and interest you can derive from observing. This way, too, you can connect with the ways in which astronomical discoveries have transformed our perception of where we belong in the cosmos.

We have just made use of the fact that the Moon is predictable. This is because it is in a stable orbit around Earth. But the Moon's orbit does not describe a perfect circle through the sky. Rather, it rotates in an ellipse with two focal points, with an average distance from Earth of 384,401 km and a maximum of 406,700 km. This lunar orbit takes 27.3 Earth days, the sidereal month, and some of its details, as we shall see, are very complex to measure.

That the Moon is regular and can be predicted was one of the earliest discoveries of recorded astronomy. It appears in lunar 'diaries' compiled by priest-astronomers of Babylon in the eighth century BCE – some of their observations still survive. Later observers continued to make meticulous observations of the westward arc described by the Moon, but aspects of its behavior remained puzzling until Johannes Kepler (1571–1630) made sense of planetary orbits, that of the Moon as well, by way of elliptical mathematics. His *Astronomia Nova* ("The New Astronomy") of 1609 devotes a quarter of its length to exploring and explaining the ellipse model, which is now expressed in Kepler's first law of planetary motion – all planets move in ellipses, with the Sun as one focus. This law achieved precision in the 1687 *Principia Mathematica* of Isaac Newton (1642–1727). Newton explained: "The irregularity of the Moon's motion hath been all along the just Complaint of Astronomers; and indeed I have always looked upon it as a great Misfortune that a Planet so near us as the Moon is…should have her Orbit so unaccountably various, that it is in a manner vain to depend on any calculation…though never so accurately made."

We have also made use of another basic feature of lunar science, the Moon's phases. The Moon emits no light of its own and shines by virtue of reflected sunlight. Its phases are produced by the relative angles of the Sun, the Moon and Earth between them.

One of the most beautiful consequences of this interaction between the three celestial objects is called Earthshine, 'the old Moon cradled in the young Moon's arms.' Earthshine describes the glow of the unillumined portion of the Moon dimly visible beyond the shining crescent. It is produced by sunlight reflected from Earth onto the Moon and then back to Earth again. Nothing demonstrates more clearly the intimacy of the relationship between the two bodies. As Galileo Galilei (1564–1642), whom we will meet many times in this book, expressed it in 1610: "In an equal and grateful exchange the Earth pays back to the Moon with light equal to that which she receives."

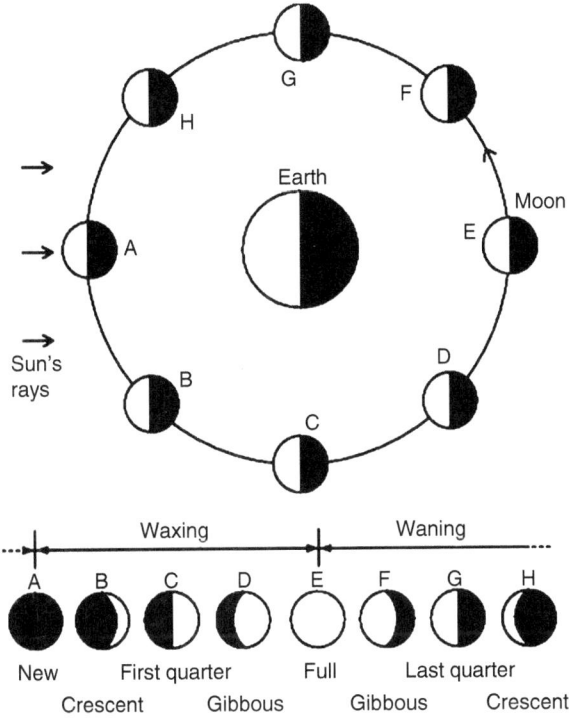

Fig. 1.4 The phases of the Moon. This simplified diagram represents the view from above the Earth's north pole (http://www.springerimages.com/Images/Physics/1-10.1007_978-3-642-14805-7_1-9)

Glimpsing Earthshine

Relatively few artists are remembered in the names given to features on the Moon, but one exception is Leonardo da Vinci (1452–1519), whose inconspicuous crater lies on the edge of the Mare Tranquilitatis. He has earned his place there, not only for being the first observer whose drawings of the Moon have survived, three in all, but also for explaining how Earthshine happened. Always interested in the way sunlight reflected off mirrors, glass and water, here are his own words, written beside the sketch of the three bodies that he drew in a manuscript now known as the *Codex Leicester:* "Some have believed that the Moon has some light of its own, but this opinion is false, for they have based it upon that glimmer visible in the middle between the horns of the new moon…this brightness at such a time being derived from our ocean and the other inland seas."

In fact, Earth's water is not a primary source of the reflected light, which comes mostly from clouds. But this is a small point compared to how Leonardo perceived the relationship between Earth and Sun and Moon. Take a look at his sketch, drawn around 1510.

Fig. 1.5 Leonardo's sketch of the crescent Moon with Earthshine, as it appears in the Codex Leicester

Da Vinci's achievement is all the more spectacular because, at the time he was drawing, prevailing orthodoxy held that Earth was the fixed center of the cosmos. The idea that Earth might be in motion around the Sun had been suggested in Greek antiquity, but in Leonardo's time this seemed almost inconceivable.

To explain how that started to change, we must return to the telescope.

Meeting Copernicus

The second half of this first observation takes us to the northern hemisphere of the Moon and another naked-eye crater, Copernicus. We can follow the same pattern as we established with Tycho and begin once again at full Moon. Under these conditions, Copernicus appears as something like a sibling to Tycho. Look further to the north of Tycho and a little 'off center.' Like with Tycho, the full Moon is the best time to notice the crater's spider's web of rays, less extended but denser than those we observed around Tycho.

It is also clear, even under these bright light conditions, that Copernicus is situated in a darker region of the lunar surface. Its surrounding terrain is flatter, and binoculars expose this as a characteristic mare landscape, in this case, the Mare

Insularum, or Sea of Islands. Through binoculars, it also becomes clear that Copernicus is an imperfect circle, more of an octagon, formed out of what seem to be separate sections of straight walls, reminiscent of a castle. Signs of landslips are visible as well as secondary craters formed from debris impact. The terraces themselves rise some 1,000 m above the surrounding terrain, and the crater itself is some 93 km wide.

The last quarter of the lunar cycle is a good time to observe Copernicus in detail. On the Moon's 24th day – an astronomical calendar or an online resource will identify this – observe with binoculars or through a small telescope the progress of nightfall up the slopes of Copernicus' central peaks until, as you watch, they vanish into the dark.

Fig. 1.6 Copernicus' wild topography has never been captured more powerfully than by NASA's Lunar Orbiter 2 on 10th November, 1966 (Credit: NASA)

The area around Copernicus is worth exploring. Just north of the crater rise the Carpathian Mountains (Montes Carptus), a fine line of peaks bounded at each end by the craters Gay-Lussac and Tobias Mayer. Continuing west from the end of the mountains and moving a little south to the level of Copernicus is the crater Kepler dug into a wild terrain of older, degraded crater rims. It presents an obviously different aspect to the flat lands of the maria terrain, an important distinction for understanding lunar processes.

The Moon Has a History

The Copernicus crater demonstrates another element of lunar science. All that decay and collapse, those multiple and overlapping craters, reveal that the Moon has a history. Some of the craters, the ones with broken walls and smaller craters within, must predate the impacts that have deformed them. Similarly, the rays emerging from the Tycho crater overlie the features all around them and therefore must have been formed after them.

Brightness is another clue. Both of our craters are substantially whiter than the gray terrain around them. But what darkens the older features of the Moon? There is no wind or rain on the Moon, so they have not 'weathered,' as on Earth, but the lunar surface has been subjected to the power of the solar wind, unmediated by any protecting atmosphere as exists on Earth. The Moon is also peppered with a hail of micrometeorites.

The area around Copernicus was the first to be studied in detail for its geological chronology. Eugene Shoemaker (1928–1997) and Robert Hackman established the relative ages of the craters, mountains and seas, in part by observing, exactly as we have just done, the overlay of ejecta from the Copernicus impact. It is now possible to identify broad periods in the lunar past of which the most recent is named, in honor of the crater, the Copernican.

The dating of the Tycho crater has been determined in two ways. The first is by a count of the smaller craters within the Tycho envelope. Older features will on average have more evidence of cratering than the younger – the older have been around to be hit for longer. Some data about the crater was also collected by NASA's Surveyor 7 unmanned lander, which touched down on the rim of Tycho in January 1968. After Apollo 17 astronauts brought back samples of Tycho's ray ejecta, it became possible to date the event precisely. The Tycho crater was formed 109 million years ago and is the youngest large feature on the lunar surface.

Giving a history to lunar features is important because, until relatively recently in human history, many people asserted that the Moon was an unblemished sphere outside the reach of change, part of a concentric system of spheres with Earth at its center. The decline of this system of thought was, a little like has happened at Copernicus, a gradual collapse, but there are definite moments when aspects of the paradigm failed. The two men remembered in these two craters played central roles in the story of reshaping the human understanding of the universe.

The Moon Is Made of Alabaster?

Medieval European astronomers were not stupid. Built upon the foundations laid by Greek philosophy, astronomy gained a new repository of insights from Islamic thinkers by way of Moslem kingdoms in southern Spain. The subject was part of the curriculum taught at medieval universities.

We get a glimpse of this astronomy in the basic textbook used throughout medieval Europe, John de Sacrobosco's *De Sphaera, On The Spheres*. Written around 1230, it was the basis for lectures and commentaries up to the end of the sixteenth century.

We know very little about John of Sacrobosco (John of Holywood). He taught in Paris and, aside from astronomy, was interested in the measurement of time. He was among the first scholars to argue for a reform of the calendar.

Sacrobosco presented astronomy as a mathematical science. He opened his work with geometric definitions of a sphere, and he went on to describe the heavens in terms of a series of spheres radiating outwards, as it were, from Earth. "The earth is held immobile in the midst of all," he wrote. This was a spherical Earth, though; the idea that all medievals thought Earth was flat is incorrect, and Sacrobosco went to some pains to prove this.

Sacrobosco explained the way the Moon, planets and stars moved through the sky by reference to more circles. It was complicated stuff:

Every planet except the sun has three circles, namely, equant, deferent, and epicycle. The equant of the moon is a circle concentric with the earth and in the plane of the ecliptic. Its deferent is an eccentric circle not in the plane of the ecliptic – nay, one half of it slants toward the north and the other toward the south – and the deferent intersects the equant in two places, and the figure of that intersection is called the dragon.

Dragons aside, this is an attempt to explain the observable facts with reference to interlocking circles, something Sacrobosco derived from the Claudius Ptolemy, whose *Almagest* was written in Greek in the second century CE. The geocentric or Earth-centered view of the universe was referred to as Ptolemaic, after Ptolemy.

It's interesting to note the evidence Sacrobosco used to demonstrate his points. It was almost all from classical authors of the Roman and Greek periods, not astronomers but historians and poets. This gives a clue to the kind of authorities that underpinned his astronomy. He wanted the support of antiquity much more than observation.

It was the same with the theory of the perfect Moon. It was never based on observation but drawn from principles espoused by Aristotle. It later acquired a significance in Roman Catholic theology as a metaphor for the Immaculate Conception of the Virgin Mary.

There were always two problems with this account. The first was obvious – the Moon had some areas that were darker than others. The second was a piece of scientific reasoning. If the Moon was perfect, then as a spherical mirror it would presumably act like other mirrors and reflect the Sun as light is caught on a billiard ball, showing one brilliant spot. This was not the case. It led observers to wonder what the Moon was made of and whether there was a property on its surface, however perfect, that led to this effect. Some wondered if the Moon might be made of alabaster, and this became the accepted explanation by the thirteenth century. As Albert of Saxony (c. 1320–1390), another significant figure in medieval astronomy, explained: "The portions of alabaster that are very dense and non-transparent appear very white; those that are transparent like glass are obscure and tend towards black. If one asks why the Moon exhibits such differences between its various parts, one must reply that this is its nature."

New Ways of Thinking

The Ptolemaic system lasted. Commentators wrote commentaries upon commentaries and on it went. Anyone who argued differently, whether for a Moon that displayed evidence of change or for a Solar System where everything, including Earth, revolved around the Sun, was calling for a revolution in thought. And not just thought. Upheavals in astronomy touched upon theology and astrology, threatening to undermine the sense of where human beings fitted into the cosmos. This was not a small thing to undertake.

We will meet many of those who did this, though, during the course of this book. Here, we will touch upon two. They are the astronomers whose names are attached to the two craters we have observed, Nicolas Copernicus (1473–1543) and Tycho Brahe (1546–1601). Both offered different models for an alternative reality.

Copernicus' universe is contained in two books, his *Commentariolus* ("Little Commentary"), written before 1514, and then the more famous *De Revolutionibus Orbium Coelestium* ("On the Revolutions of the Celestial Spheres"), published in 1543. In these books, Copernicus argued historically, framing his new model in the familiar and comforting terms of what came before him. "To be sure," he wrote in his preface dedicated to Pope Paul III, "Claudius Ptolemy…brought this entire art almost to perfection…. Nevertheless very many things, as we perceive, do not agree with the conclusions which ought to follow from his system." In place of the complex Ptolemaic system, Copernicus proposed something simple, one system applying to all. He proposed that Earth and the planets revolved around the Sun because "the Sun occupies the middle of the Universe."

Like Sacrobosco, Copernicus' used a mathematical argument. Copernicus' title page carried the warning, "Let no one untrained in geometry enter here," said to have been the motto inscribed over the gateway to Plato's academy in ancient Greece. The text lives up to this, punctuated with tables and diagrams of interlocking triangles. It was a careful, almost reluctant piece of work.

Copernicus' publications did not immediately dissolve the Ptolemaic consensus. But his works were read, and some of those who thought it through came to agree with what he was saying. This, for example, is Michael Maestlin (1550–1631), tutor and mentor to Johannes Kepler, writing in the margin of his copy of Copernicus' *De Revolutionibus*: "Unless the common hypotheses are reformed…I will accept the hypotheses and opinion of Copernicus." Maestlin would live to see the gathering of evidence that Copernicus was right.

Others thought through the problems of the Ptolemaic system differently. One of the greatest astronomers to turn over the evidence was Tycho Brahe. In his own view he was a prince among observational astronomers, and he worked from the Danish island of Hven in the Baltic Sea to amass vast quantities of data with every kind of instrument he could create. He used this to propose a different model.

In the Tychonic account, the Moon orbited Earth and so did the Sun, with the other planets then in orbit around the Sun. This geo-heliocentric model made sense of his voluminous observations, on the basis of which, in Tycho's view, "the position of Copernicus on the motion of Earth and the immobility of the Sun is weakened." Tycho was a giant among astronomers, and his model gained support during the seventeenth century.

Return now to the craters that carry these two names. There, on the surface of the Moon, cartographers recalled two systems vying for prominence over a period of revolution in astronomical thought. The outcome was never obvious, and there are also astronomers remembered on the lunar surface who supported the Ptolemaic system and the Aristotelian philosophy that underpinned it. What we take as self-evident is not so until you look. The telescope has changed the way we see the universe, so use it carefully. It may change you.

Chapter 2

How to See the Sun

Observation: The Sun, the solar corona, auroras
Significance: Astronomy beyond the visible, the rotating Sun, solar wind
Science: How the Sun works, sunspots

The Sun is a star, and all stars are suns. The word *Sun* is the name we give to our local star. Its first recorded use in English comes from the hand of King Alfred in the ninth century, but most Indo-European languages have similar-sounding words for it based around a root syllable *su-* or *so-*. We have been talking about the Sun for a long time.

The Sun is, right now, the most important thing in your life. Its power created the Solar System, and its gravity holds it together. It is far and away the largest astronomical body in our local environment. Just 150 million kilometers away and 1.4 million kilometers across it dominates the daytime skies. With the correct apparatus, it reveals an awesome glory. We can watch its power at work.

In this chapter we will, after taking due precautions, observe the Sun. We will also see how those observations have changed history. It was upon the face of the Sun that astronomers saw blemishes, spots that overturned another element of the ancient view of the heavens and demonstrated that the Sun was in motion. It was by dissecting sunlight that astronomers first stretched the boundaries of their science beyond the visible. It has been by identifying and studying the solar wind that we have been able to understand our Solar System as the realm where the Sun's power rules.

These have mostly been quiet revolutions. When Galileo saw the moons of Jupiter, it was obvious to many that the classical account of the sky was incomplete. The same was true for sunspots, which provoked a furious debate in the seventeenth century. But spectroscopy, the dividing of light, and the nature of the solar wind are huge advances in the human understanding of our world that have remained hidden. Seeing the Sun is a chance to make them shine.

M. Marett-Crosby, *Twenty-Five Astronomical Observations That Changed the World: And How To Make Them Yourself*, The Patrick Moore Practical Astronomy Series, DOI 10.1007/978-1-4614-6800-4_2, © Springer Science+Business Media New York 2013

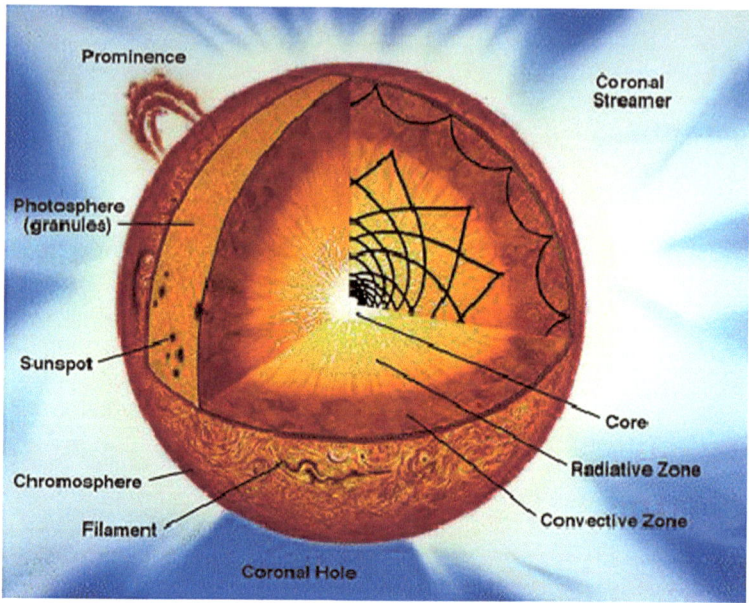

Fig. 2.1 A cutaway diagram showing the structure and zones of the Sun (Credit: NASA)

How to Look at the Sun

Here is a basic rule – don't look at the Sun.

To put it more precisely – don't observe the Sun using any telescope, binoculars or other optical instrument that has not been fitted with a safe appliance that is specific to solar astronomy. Don't stare at the Sun with your unaided eyes, unless you are using a recognized and tested solar filter. Why? Because human sight is precious and fragile, and the Sun can cause immediate and irreversible damage to the eyes.

This means avoiding home-made pieces of smoked glass, color film or whatever else seems like a clever idea and isn't. This means never using photographic solar filters as a sunshield for the eyes because, as their name suggests, they are for cameras.

This means, above all, not taking risks. If you are unsure of a piece of equipment, then do not use it to look at the Sun.

If you observe these warnings, then solar observing will become one of the pleasures of astronomy. Daytime viewing can be plagued by weather, but the Sun is much more available than any other astronomical target. It is also changing all the time, and we can see that for ourselves. This is rare in astronomy, where the times-cales are usually far longer than the span of many human lives. The Sun provides a chance to understand and use fantastic online data provided by satellites that are watching it all the time. Finally, learning about the Sun informs our observations of other, much more distant stars.

In this chapter, we are going to explore the Sun using each of the safe ways open to amateur astronomers. We will examine white light, hydrogen alpha, coronal and auroral observing, looking first at how to make the observation and then at what it reveals. This way, we will build up a practical picture of what we know and also what we don't know. The bright Sun reveals everything else, but hides itself.

Looking at Sunlight

In 1800, the great astronomer William Herschel (1738–1822), whose work with his sister Caroline (1750–1848) brought about huge changes in our understanding of the Solar System, used a prism and a thermometer to show that the Sun's heat extended beyond visible light. It was a simple experiment, easy to repeat at home. It was a small step in the wider search for the nature of light, a demonstration that the visible was not all that light contained.

The modern analysis of starlight began with Isaac Newton (1642–1727), a towering figure in the history of science to whom we will return in later chapters. He observed in 1666 that sunlight could be divided into a continuous spectrum, the rainbow. It was Joseph Fraunhofer (1787–1826) who identified dark gaps in the Sun's spectrum. Working at his optical factory near Munich, he used a 25 mm telescope to observe (with care) the Sun, writing: "I found with this telescope almost countless strong and weak vertical lines."

Looking at white light divided into its colors, Fraunhofer identified 10 of these strong lines and as many as 570 faint ones. In 1814, he was able to draw some 350 of them in all.

Later, he extended this search to stars beyond the Sun. He looked at Betelgeuse, the red alpha star of Orion, finding there yet more "countless fixed lines which, with a good atmosphere, are sharply defined." Even more significant was what he found in the brightest star of all, Sirius: "I have seen with certainty in the spectrum of Sirius broad bands which appear to have no connection with those of sunlight."

The Fraunhofer lines found in the Sun were studied intensively over the nineteenth century. Gustav Kirchhoff (1824–1887) identified that one of the bands was related to the element sodium. He later produced his laws of spectroscopy that distinguished between the different kinds of spectrums.

Spectroscopy allows us to peer into the chemical contents of the Sun and stars. It enabled the discovery that unlocked the secret of the Sun's power.

Finding Helium

A quick glance at the Periodic Table shows two elements at its peaks. One is hydrogen, the most prevalent element in the universe. The other is helium, identified in the nineteenth century by looking at Fraunhofer lines in the Sun.

The discovery of helium is credited jointly to Pierre Jules Janssen (1824–1907) and Norman Lockyer (1836–1920). Janssen made the journey from France to Andhra Pradesh in India to observe the solar corona, and he identified a dark absorption line never seen before. He thought it might be an unknown element lodged within the Sun and sent a communication to this effect to the Academy of Sciences in Paris.

By coincidence, the same October 1868 meeting of the Academy had a report announcing the same discovery from the British astronomer Norman Lockyer. Janssen might have saved himself the trip, for Lockyer had identified the unknown absorption line while observing from West Hampstead in London. The new line was quite near two familiar lines, D1 and D2, the fingerprints of sodium, but it could not be reproduced in the laboratory. Lockyer also proposed that the D3 line represented a new element.

This was a controversial suggestion. New elements with names such as asterium and coronium were being proposed in the scientific literature of the time, and one alternative theory was that D3 represented hydrogen but under extreme pressure and temperature. The existence of what became known as helium was not accepted until 1895, when William Ramsay and Morris Travers isolated it from gases given off by uranium minerals.

Lockyer also established the journal *Nature*. The first issue stated that it aimed to "place before the general public the grand results of scientific work and scientific discovery; and to urge the claims of science to move to a more general recognition in education and in daily life." This and his other many achievements led to his being gently lampooned:

And Lockyer and Lockyer
gets cockier and cockier;
For he thinks he's the owner
of the solar corona.

The White Light of the Sun

What makes the Sun shine? The Sun is in one sense insubstantial, a luminous ball of plasma without any solid surface. It is concentrated and very dense at the center, where the heat and pressure are so great that atoms of hydrogen collide with one another in a process known as nucleosynthesis or, more familiarly, fusion. First proposed in the 1920s by Sir Arthur Eddington (1882–1944), it was Hans Albrecht Bethe (1906–2005) who established a sequence of chemical reactions taking place to convert hydrogen protons via several stages into helium, the Proton-Proton (PP) chain. This releases a huge amount of energy that flows outward from the Sun's core to the edge and emerges as visible light. This white light of the Sun is what we see every day. It is white because it contains all the colors of the spectrum.

"Direct the telescope upon the Sun as if you wanted to observe it," wrote Galileo. "The further the paper is removed from the tube, the larger the image will become and the better the spots will be depicted, and without any injury one will see all of them."

It is possible to observe the white light of the Sun. The warnings with which this chapter started are serious, though, and no observation of the Sun should begin without taking all proper precautions. The Sun in white light can be safely observed with either a dedicated white light solar telescope or a safety-tested commercial (not home-made) adapter attached to a nighttime instrument, or by projecting the Sun's image onto a screen. Once again it must be done with care – care for the telescope, care for the screen and above all care for those observing. Never look directly at the Sun through a telescope's eyepiece.

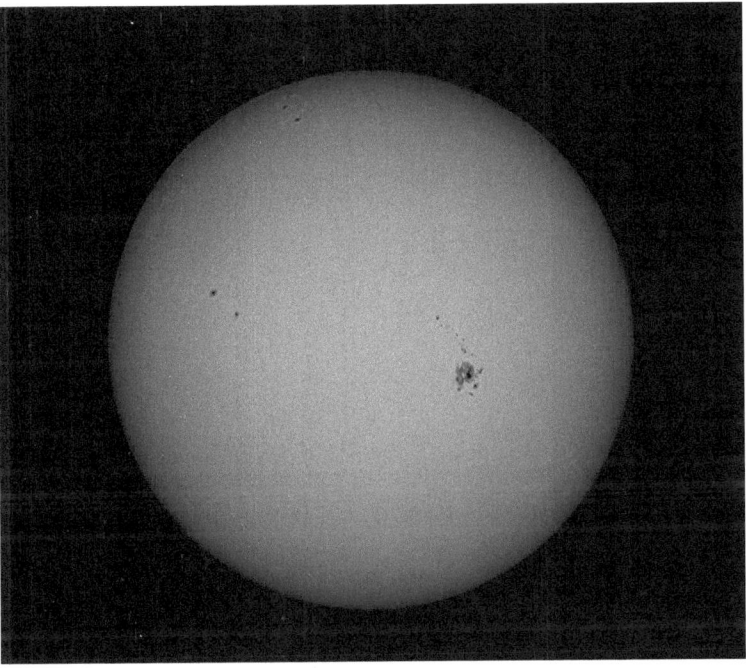

Fig. 2.2 The Sun in white light on May 12, 2012, taken with Baader solar film (density 5.0) with Canon 40D fitted with a Baader Solar Continuum filter, all through a Celestron ED80 Scope mounted on an Astrotrac (Credit: Stephen Devine, www.urban-astronomy.com)

The image produced by these methods reveals something like this. We are observing the solar photosphere, a white-gray environment punctured, especially in latitudes towards the Sun's equator, by the distinctive dark marks of sunspots.

These are the outer signs of intense and violent magnetic activity within the solar mass. They are areas of reduced temperature, a 'mere' 3,000 or so degrees of heat, but often many thousands of kilometers across. Sunspots begin life in latitudes to the north and south of the solar equator – we'll assume here that north and south represent compass points analogous to those on Earth – in activity belts where they are the product of a complex interplay of magnetic forces working through the various different rotations on the surface and within. Different areas of the Sun rotate at different speeds, creating twists and buckles in the magnetic fields. A sunspot probably comprises twisted rope structures wherein these magnetic forces are impeding energy from reaching the surface. Sunspots always move towards the center of the solar disk, changing all the time.

As you observe sunspots safely, you might start to notice light bridges that come to divide the dark umbra of a sunspot into two and also faculae (singular, *facula*), bright points near a sunspot where energy is escaping.

At solar minimum, sunspots form around, and rarely beyond, latitudes of 45 °. As the Sun moves towards solar maximum, the period of its peak magnetic activity, the sunspots move, as we observe, towards the center. This waxing of solar activity lasts 11 years and the full cycle of sunspot activity – the Schwabe Cycle, identified by Samuel Heinrich Schawbe (1789–1873) and recorded by Rudolf Wolf (1816–1893) – is therefore completed in 22 years.

White light observing reveals the power of the solar cycle. If you look at or near a well-developed sunspot, taking the precautions indicated above, you may be fortunate enough to observe a patch of temporary brilliance, perhaps half as bright again as the rest of the photosphere. These are white light flares – D, E and F-type sunspots as classified in the Zurich-McIntosh system are especially prone to them. They are brief but spectacular markers of the forces within.

A Funeral for Pseudo-Philosophy

Sunspots matter. First seen by Chinese observers, their existence led to one of the fundamental debates of early modern astronomy. It was a battle over how the universe, and science, should work.

On one side was the Jesuit priest-scientist Christoph Scheiner (c.1573–1650), who in March 1611 made use of a heavy mist to observe dark spots upon the bright orb. He returned to the same observation later in the year, shielding himself from the solar glare using stained glass. While his methods are not recommended, his observations were important. He wrote three letters describing his discoveries and had no doubt what he was seeing. "I have always considered it inconvenient to place spots… on the bright body," he states, before advancing his argument that they were bodies transiting in front of the Sun, little planets like Mercury or Venus:

> "I do not think, therefore, that they are real spots, but rather bodies partly eclipsing the sun, namely stars located either between the sun and ourselves or revolving around the Sun."

This conclusion freed the Sun from blemishes. Scheiner was protecting the idea that the heavens were both perfect and unchanging. "We saw what we were looking for," Scheiner wrote tellingly a little later – all was well with the cosmos.

Galileo disagreed. He, too, was observing sunspots, and in a letter of 1612 he announced his intention to debate them, for: "Sunspots should bring about the funeral, or rather the extreme and last judgment, of pseudo-philosophy."

Galileo had no time for Scheiner's views. "Continued daily observations show me," he wrote, "with every conceivable confirmation and no contradiction whatsoever, that my opinion squares with the facts." These were, he argued, spots that resided on the surface of the Sun. His was a case based upon observation and also geometry. It was the simple solution, requiring no intermediate unknown bodies. "The spots are contiguous to the surface of the sun, and are carried around by its rotation."

Sunspots revealed an imperfect Sun. They also showed that the Sun was itself in motion. It is exciting still to observe the motion over time of sunspots, but in the seventeenth century it was to effect a revolution. Gone were the immobile upper spheres of the sky. In place of a philosophy hallowed by antiquity and authority, Galileo was proposing the primacy of observation. Scheiner had seen what he wanted to see. Galileo had just seen.

The Boiling Sun

Our second solar observation takes us beyond the experience of Galileo and Scheiner and into the chromosphere.

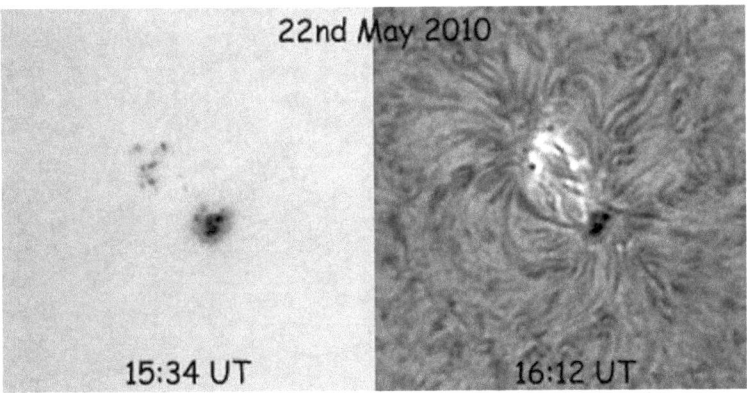

Fig. 2.3 A sunspot compared. Images of AR1072 in white light (*left*) and hydrogen alpha (*right*) from 22 May, 2010 with an 80 mm refractor and Baader solar filter for the WL image and a Coronado PST for the hydrogen alpha together with an Imaging Source monochrome DMK AU03 camera, images processed using RegiStax (Credit: Peter Meadows, www.petermeadows.com)

A hydrogen alpha telescope enables observers to explore the solar chromosphere, which lies above the white light photosphere. This is a zone both more tenuous and more transparent than that which we observe at white light, some 2,000 km thick and with an average temperature of around 10,000 K. The chromosphere contains no permanent features, its mottled patterns forever changing as convective currents rise and peak and fall. It's in the chromosphere where observers spot solar flares, sudden and massive releases of the energy stored in the Sun's magnetic field.

Hydrogen alpha telescopes use etalons, devices with parallel reflecting plates that reflect off unwanted wavelengths and, in their modern form, eliminate more than 99 % of the Sun's light. The name etalon was applied to the device by its creators, Charles Fabrey and Alfred Perot, in 1897 (Fabrey was later a co-discoverer of the ozone layer) and is taken from the French *étalon*, literally meaning *standard*, as in a standard unit of measurement. The other English word derived from *étalon* is stallion, the stallion being, I suppose, a standard horse. It's an odd combination.

It is worth taking any opportunity to observe at this or other wavelengths that reveal the chromosphere. There are prominences, which are graceful, almost balletic in the way they curl. Something about them inspired the Menzel-Evans scheme that labels features in this zone to abandon dry codes in favor of descriptive nouns. You can spot hedgerows or coronal rain, funnels, loops, surges and puffs. There are even jets and tornadoes.

The Wind in Our Hair

Perhaps the greatest revelation of solar power, though, comes when we have an opportunity to observe the Sun's corona.

The first way that this can be done safely is during a solar eclipse. The Moon, perfectly in line between the Sun and Earth, shields us from the white light photosphere such that it becomes possible to see the corona's streamers and rays. But total eclipses are rare events, and so in the early 1930s the French astronomer Bernard Lyot created an instrument to impose an artificial moon in between the Sun

Fig. 2.4 The solar eclipse of August 1, 2008, at the point of totality, when the moon completely blocks out the body of the sun, revealing the corona (Credit: The Exploratorium, NASA)

and the observer. The coronagraph has enabled astronomers to study the outer area of the Sun, just one millionth of the brightness of the main photosphere.

What we are observing here is solar gas flowing into space. It was thought for a long time that the incredible temperatures of the corona and its corresponding Fraunhofer lines were evidence of a special chemical element. In fact, these lines show familiar terrestrial elements but under extraordinary conditions. The corona is expanding in every direction, creating a wind of charged particles. If you use a coronagraph, you will see streamers trapped by solar magnetism flow back towards the Sun in helmet loops while others spread into the darkness of space.

Where do they go? Observing the solar corona is a good place from which to glimpse the solar wind.

The existence of this wind is now so much part of our understanding of the Solar System that it is easy to forget how recent is its discovery. It is also a good example of how theory and observation interact, and sometimes don't, in science.

The earliest ideas that there was some sort of flowing solar force came from observing how comet tails always point away from the Sun. This led astronomers to conclude that something was pushing the tails to do this, perhaps sunlight itself.

The idea of solar wind emerged out of experiments conducted by the Norwegian Kristian Birkeland (1867–1917). Writing a report on the Norwegian polar expedition of 1902–3, he proposed that "aurora and magnetic perturbations should be regarded as rather moderate manifestations of an unknown cosmic agent of solar origin." Conflicting models of the way the solar corona might work were constructed later by Ludwig Biermann (1907–1986) and Sydney Chapman (1888–1970), Biermann proposing a stream of particles he called corpuscular radiation, and Chapman arguing for a corona that was static and not expanding. In 1958, Eugene Parker entered the fray, creating the phrase 'solar wind' to describe the relentless outward expansion of the Sun.

If the theory of the solar wind was right, could it be measured in practice? By 1958 both the United States and the USSR were sending satellites into space, but measuring the solar wind proved far from easy. Three Soviet satellites tried to feel the wind in 1959, and they did provide some data in support of Parker's theory. So did the Soviet Venus probe of 1961. Meanwhile the American *Explorer 12* satellite seemed to prove the negative during a 4-month mission in the same year. The next three U. S. satellites with experiments on them all failed. But the overheating, battered U. S. spacecraft *Mariner 2* ended the uncertainty in 1962. There was a solar wind, and it was blowing past *Mariner 2* in a steady gale.

It turned out to be a gale of two parts. One of the puzzling and fascinating results from the study was the discovery of the two phases of the solar wind, the fast and the slow. Like all the best experiments, then, *Mariner 2* presented answers and then asked the next set of questions.

Recent satellites have taught us more about the solar wind. The joint NASA-ESA *Ulysses* spacecraft, which ended operations in 2009, provided data from above the northern and southern poles of the Sun and studied variations in the solar wind during maximum and minimum periods of solar activity. NASA's ACE

(Advanced Composition Explorer) satellite is contentedly sampling and studying the solar wind and will continue doing so until 2024 if its fuel holds out. Such are the advances in our ability to measure the solar wind that it is now possible to get a daily reading of the solar wind on a personal computer.

The *Skylab* space station carried eight solar experiments, and X-ray images taken during the manned missions of 1973–1974 enabled a first prolonged view of the Sun's hot corona. Skylab's manned missions ended in 1974 and plans to revisit it were curtailed by the same Sun that the space station had been so assiduously studying. Predictions of solar weather had failed to predict a burst in the Sun's activity that had increased the drag on the spacecraft, pulling it out of position. It was this that led to the early demise of the first U. S. space station, brought down by the Sun.

In 2001, NASA launched a remarkable attempt not only to measure but also to bring back to Earth a sample of the solar wind. The *Genesis* spacecraft was the first sample-return mission since *Apollo* and spent 886 days collecting the wind's diffuse ions. The return capsule entered Earth's atmosphere on September 8, 2004, but a design error resulted in the parachute failing to deploy, and the precious cargo crashed into the Utah desert. Amazingly, it proved possible to rescue the solar wind collector from the debris.

From this and other evidence, it now seems that the oxygen and nitrogen isotopes in the solar wind are like those found in the atmosphere of Jupiter but unlike those common on Earth. We don't yet know what this really means. Studies of the solar wind, and our place within it, may yet reveal many secrets of the power and influence of the Sun.

Just how powerful is the Sun? All the planets are embedded in the solar wind, but it is of course stronger nearer the Sun than in the outer reaches of the Solar System. At Mercury, 0.39 AU from the Sun, the solar wind's density is 53 cm^{-3} and around Earth 7 cm^{-3}. By Jupiter, it has diminished to 0.2 cm^{-3}, and by distant Neptune, it amounts to just .0006 cm^{-3}. Beyond Neptune, the Sun's power tails off into an area called the heliopause, where the *Voyager* spacecraft are now at work.

The role of spacecraft in measuring the solar wind is only one part of the array of remote probes doing what we cannot do safely, staring at the Sun.

2012/05/28 01:00

Fig. 2.5 This image was taken by NASA's SoHo mission on day of writing (Credit: NASA/ ESA)

Their data is readily available on the Internet, such that it is perfectly normal today to be informed of current solar conditions and download extraordinary images of solar activity. Websites for NASA's SoHo (Solar and Heliospheric Observatory) and SDO (Solar Dynamics Observatory) provide images of the Sun for each day, such as the one above, as well as movies and special reports on prominences and sunspots. These and other data mean that it's possible to check observations, particularly of sunspots, against the evidence of these immensely powerful instruments.

Sound and Glory

The images of the Sun captured by these spacecraft are awe-inspiring. But nothing binds science to wonder so closely as the Northern and Southern Lights. They reveal the solar wind.

Humans have been recording aurorae since the Babylonians made astronomical notes on baked clay tablets in the sixth century BCE, and they remain among the most beautiful revelations of how we are bound within the solar environment.

What causes the aurorae? As solar particles move into Earth's magnetosphere they travel to its day or night side along the magnetic field lines. When these magnetic field lines reconnect in an area known as the magnetotail, energy is released, and this sends the particles down onto Earth's poles, and sometimes even lower latitudes. As the particles bombard oxygen and nitrogen in the upper atmosphere, the atoms release photons of light that we see as the auroral colors.

The details of the science remain uncertain. In 2007, NASA launched the Themis mission to study the interaction between the solar wind and Earth's magnetosphere. The mission hopes to identify which of the several theories connecting solar storms with aurorae might be correct. At time of writing, Themis's mission has been extended. It could help us unlock the aurora's secrets. But it will never take away the wonder of the Sun's power at work.

It is also possible to listen to that power. Not only do some people claim to be able to hear the aurorae, but violins of the late seventeenth century, known for their superior quality, probably derive their sound from the power, or lack of power, in the Sun. It is thought that great Stradivarius instruments derive their timber from the particular strength and density of the wood out of which they were made. The trees from which the materials came grew during a period of exceptional solar quiescence at that time – known as the Maunder Minimum.

It seems likely that the unique conditions created by the Sun's half-century doze are the cause of some of the world's sweetest music.

Chapter 3

How Stars Work

Observation: Ursa Major stars, splitting Mizar and Alcor
Significance: Recognizing the variety of stars
Science: Spectral types and magnitudes

This chapter opens a longer and lonelier journey than that to the Moon or the Sun – we will begin to observe the stars. We are going to spend time with the constellation Ursa Major, moving in and out of doors between what we can see and the science that explains the observations. The discovery that stars were not uniform but existed in many states is the bedrock that sustains our understanding of the universe. We will explore something of the human processes that created the modern constellations, and conclude by meeting someone who embodies the muddled history we have exported onto the heavens.

Finding the Dipper

He who would scan the figured skies
Their brightest gems to tell
Must first direct his mind's eye north
And learn the Bear's stars well.

It might not be great poetry, but these lines by the American amateur astronomer William Tyler Olcott (1873–1936) in his book *Star Lore of All Ages* (1911) make their point. The constellation Ursa Major, the Great Bear, is a feast. Naked eye, binocular and telescopic observing each uncover some part of its wonder. At its center lies a seven-star asterism, not a constellation but a familiar shape within it, the Big

M. Marett-Crosby, *Twenty-Five Astronomical Observations That Changed the World: And How To Make Them Yourself*, The Patrick Moore Practical Astronomy Series, DOI 10.1007/978-1-4614-6800-4_3, © Springer Science+Business Media New York 2013

Dipper or Plough. Four stars make a bowl or ladle, three a crooked handle leading from it.

Fig. 3.1 Ursa Major and the Big Dipper asterism with the constellation Coma Berenices in the *lower left*, wide-field view (Credit: A. Fujii)

If you can't see this, then the chances are that your seeing is diminished by overmuch ambient light. In an urban or suburban setting, the spread of streetlamps blinds us to the skies. There are some things we can do to protect our precious night vision – an obvious but sometimes overlooked one is to switch off our own house lights, another is to use a red rather than a white beam torch when working around the telescope – but against the curse of city glare, there is not much else to do but start from somewhere else, from which we might be able to say something like, "A more advantageous, and may I say, pleasant place for astronomical observations it would be scarcely possible to find."

So wrote Charles Pritchard (1808–1893), who in 1883 traveled to Egypt, so as to observe the skies. His chosen site was Abbaseeyeh, 3 miles from Cairo on the edge of the desert. There he found what he was looking for, a place with no lights that interfered with measuring the relative brightness of some 2,000 stars. The results were published in 1885 in his book *Uranometria Nova Oxoniensis*.

This represents an extreme solution, but dark sky centers and even dark sky islands are now being marketed to meet the challenge of the light. They are a response to a phenomenon of the modern age. For the first time in human history, we are losing our sense of the stars.

Various seeing scales now exist, providing a practical way of setting realistic observing goals. They are also vital if we want to undertake any systematic recording of delicate phenomena, whether that be the cloud patterns of Jupiter or Saturn or the features on Mars. There are various such scales, including the Antoniadi and one created by the Association of Lunar and Planetary Observers (ALPO).

Here is a simplified ALPO scale:

1	Very poor images, impossible to see details or to sketch
2–3	Almost continuous distortion with occasional brief good moments
4–6	More continuous distortions with short intervals of good seeing
7–8	Intervals of perfect seeing with fine scale distortions between
9–10	Perfect seeing with steady images at high magnification

Find the star that lies at the far end of the Big Dipper's handle, where the asterism comes to an end. It is called Alkaid, an Arabic word meaning *leader*, and has the Bayer classification Eta Ursae Majoris (we will unpick what this kind of label means in another chapter). It is going to be our first observation, and around it we will gather some of the science of the stars that will underpin much else in this book. Capture Alkaid in binoculars, and it will shine blue-white.

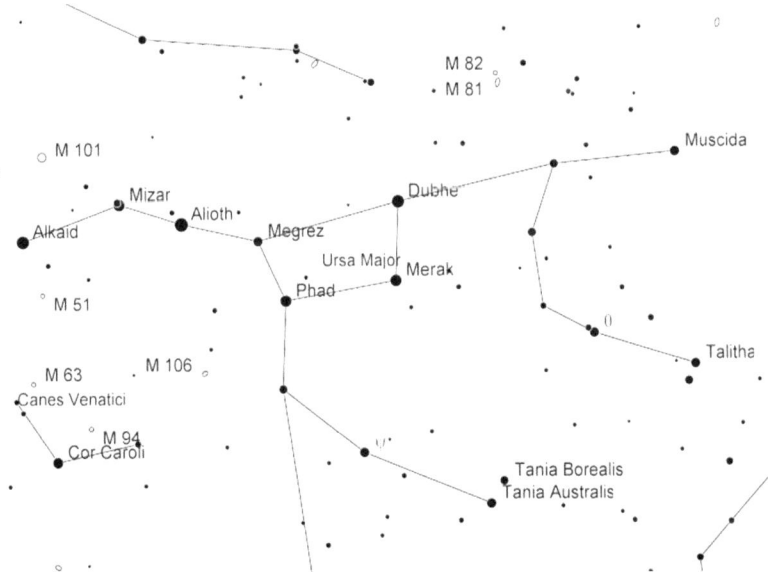

Fig. 3.2 The central portion of Constellation Ursa Major (Cartes du Ciel)

Compare this with the other stars in the asterism. You will encounter different hues, from white and blue to twinkling gold. This is not merely aesthetics. Color is a measure of stellar temperature. The order is slightly contrary to instinct: red is the coolest color and blue the hottest. These provide the basis for a scheme of classification, which in its simplest form distinguishes seven groups of stars. The full scheme contains more, but for the present we will focus on those contained in the mnemonic *Oh be a fine girl* (or *guy*) *kiss me.*

Class	Temperature (Kelvin)	Color
O	31,000–50,000	Blue
B	10,000–31,000	Blue to blue white
A	7,500–10,000	White
F	6,000–500	Yellowish white
G	5,300–6,000	Yellow
K	3,800–5,300	Orange
M	2,100–3,800	Orange-red

The process by which we arrived at this system was complex, but arose out of the efforts of astronomers – William Huggins (1824–1910), Father Angelo Secchi (1818–1878), E. C. Pickering (1846–1919) and especially Antonia Maury (1866–1952) and Annie Jump Cannon (1863–1941) – to make sense of the bewildering variety of spectra extracted from the light of stars. We will meet all of these figures again later. The above system is a synthesis of their and other schemes, and it is not stable yet. New letters have been added, and there may be more to come. The system enables us to put like stars together, but the letters embrace a lot of variety as well. In particular, a single letter will contain both giant stars, main sequence stars and dwarfs – these are categories we will look at again in another chapter.

There is a further simple but vital observation that can be made with Alkaid. Notice its brightness relative to the other stars in the asterism. Brightness is a basic unit of astronomical measurement.

The scale of brightness we use today is founded upon observations made over two millennia ago by Hipparchus (around 129 BCE) and Claudius Ptolemy (around 140 CE), who between them established a basic structure to what they termed magnitude. In their early catalogs of stars, they distinguished between stars of the first magnitude and stars of lesser magnitudes down to six, making six, therefore, less bright than one. Since Ptolemy's list, his *Almagest,* remained the basic textbook for astronomy for the next 1,400 years, the Hipparchus-Ptolemy system stuck, refined and perpetuated by medieval observers such as the Persian al-Sûfi (903–986), who subdivided the Ptolemaic values with 'less than' and 'more than' to match his more precise observations.

It was only with telescopes that this system started to creak. Galileo reported seeing stars far fainter than his predecessors had glimpsed, "such a crowd of others that escape natural sight that it is hardly believable." This crowd's magnitude he set, logically enough, at seven.

This rough-and-ready system was easy to understand. But it was observational rather than mathematical, and if magnitude was going to have meaning in modern astronomy, it needed structure and number. It needed to mean something calculable.

Enter Sir Norman Pogson (1829–1891), the astronomer who established a scale whereby one level of magnitude corresponded to a brightness difference of the fifth square root of 100. This enabled magnitude to be measured on a logarithmic scale, and the Pogson ratio made sense of the numbers.

However, astronomical discovery was outpacing the scale. The ancient system of backwards counting, now buttressed by Pogson's logarithms, had to cope with stars a lot brighter than Hipparchus's first level of magnitude and had nowhere else to go but into negative numbers. Thus the Sun, the brightest object we see, has a magnitude of −26.7, Sirius scores −1.4 and Vega, the fifth brightest star in the sky, has a magnitude of 0. Values like these feel contrary to fact, but magnitude is a good example of the interaction in astronomy between observation, mathematics and history. No doubt if we were to start afresh, we would not measure like this. But whenever we look into the sky, we follow pathways made by many other eyes. Hipparchus, Galileo and Sir Norman Pogson have all been there before us.

Astronomers now distinguish between two types of magnitude. The first, apparent magnitude, is exactly that which Hipparchus measured, how bright something is as seen from Earth. Absolute magnitude, by contrast, is a measurement of a star's luminosity as it would be at a set distance of 10 parsecs for stars (about 32.616 light years) or 1 AU for Solar System bodies, assuming there is nothing blocking the view. Both systems of magnitude still use the ancient and modern scale.

Let us now locate Alkaid in these systems. The star belongs to the B spectral group. This makes it unlike the rest of the asterism – five of the stars are A-type and Dubhe, the front lower bowl star, is from the K group. The family of B stars includes some of the brightest in the night sky. They range between very luminous supergiants and much fainter dwarf stars. Alkaid has an apparent visual magnitude of +1.84. It is in fact one of the so-called anchor points guiding the classifications of other stars.

Horses and Riders

We have used the first star of the Big Dipper asterism to acquire a set of tools with which to explore the stars. We can make use of this and expand upon it by looking at the next star along the line of the handle. In good seeing conditions, you might be able to glimpse that it is not one star at all but two, and a quick look through binoculars will confirm this. The larger star is Mizar (Zeta Ursae Majoris), and its smaller companion a little above and to the left is Alcor. These are sometimes referred to as the Horse and the Rider, although their Arabic names do not translate to mean this.

Mizar was separated first not by Riccioli, as is sometimes claimed, but in the circle of astronomers around Galileo. The Benedictine monk Benedetto Castelli (1578–1643) urged Galileo to look at Mizar in January 1617, saying "It is one of the beautiful things in the sky and I do not believe that in our pursuit one could desire better." Inspired by this, Galileo made detailed observations of the star – his handwritten observing notes survive.

For many centuries, it was a quick and easy test of eyesight – could you split the stars? They have always drawn observers. And the more you look, the more you see. Examine Mizar through a small telescope and you will split this star into two components, Mizar A and Mizar B. These are a pair of hot diamonds, two stars of spectral type A first separated at the dawn of the age of the telescope.

Mizar and Alcor are close to one another, but do they interact? For a long time, we were not sure. Separated by some 3 light years they are traveling at comparable velocities, indicating that they started from a single beginning, and data gathered in 2009 does seem to demonstrate that Alcor is bound to Mizar with an orbital period of some 75,000 years. It is worth remembering this number and then renewing the observation – the tiny gap we can resolve between the stars represents this vast stretch of distance and time.

If you place Mizar in the center of the telescope and then identify Alcor, you will see between them the faint star Sidus Ludovicianum or Stella Ludoviciana, Ludwig's Star, so named in the 1720s by Johann Liebknecht (1679–1749), who was trying to gain the favor of his monarch and thought the star might be a new planet. At a magnitude of around +7.5, it is a tiny dot compared to the Mizar pair but is of the same spectral type A as Alcor. It is fainter because it is much further away.

The wonders of Mizar have increased with each new advance in astronomical technique. Working at Harvard in 1887, Antonia Maury found that Mizar A was in itself a double star, although one that could not be separated visually. She determined this by studying the spectra of light the star produced – the lines were double, hence two stars. So Mizar A became the first spectroscopic binary, comprising two stars equal in brightness and with an orbit of just 104 days around each other. In 1908, it became clear that Mizar B was also a double star, creating from the one original a quadruple system with Alcor in attendance.

And what of Alcor? The star was imaged as part of a 2009–2010 survey of companions to nearby stars at the MMT Observatory atop Mount Hopkins in Arizona. Using its powerful Clio instrument, astronomers secured images of a small star, Alcor B, living in the glare of Alcor A.

Why does all this matter? Observations of Mizar and Alcor have expanded dramatically our sense of how stars work. They introduce us to the way that stars can relate to close companions. Amateur instruments cannot hope to replicate the achievements of Mount Hopkins or Antonia Maury, but by splitting Mizar we can share in Benedetto Castelli's first glimpse of what the telescope would bring.

There is another pair of stars to split immediately adjacent to Mizar and Alcor. The third star along the line of the handle is Epsilon Ursae Majoris, the star Alioth. If you watch it carefully through a telescope but without staring, using the edge of your eye in a useful technique called averted vision, it is possible to separate the main star at magnitude +2 from another above and to its left, the star 78 Ursae Majoris.

Epsilon Ursae Majoris is also an interesting star in itself. First of all, it varies in brightness, a feature of many stars that we will look at in a later chapter. It is also a CP, letters that stand for Chemically Peculiar. CP stars reveal unusual compositions when their spectra are examined. Some (Type Am or 1) show the presence of heavy metals and others (Ap or Type 2) strong magnetic fields combined with metals like chromium and strontium. A third group are Hg-Mg stars, containing large amounts of mercury and manganese, while a fourth are He-weak stars, displaying less helium in their mass than there should be. Alioth is an Ap Type 2 star and shows bands of separated chromium.

There has not always been a Big Dipper in the sky. Compare these three images to see the difference between today, yesterday and tomorrow. The stars are not still.

The fact is, each star has a proper motion. Stars are all on the move, each at different speeds – the record is held by Barnard's Star in the constellation Ophiuchus. Some of the stars we are observing are in motion as a group. Three of the four stars that make up the bowl of the dipper (Megrez, Phecda and Merak)

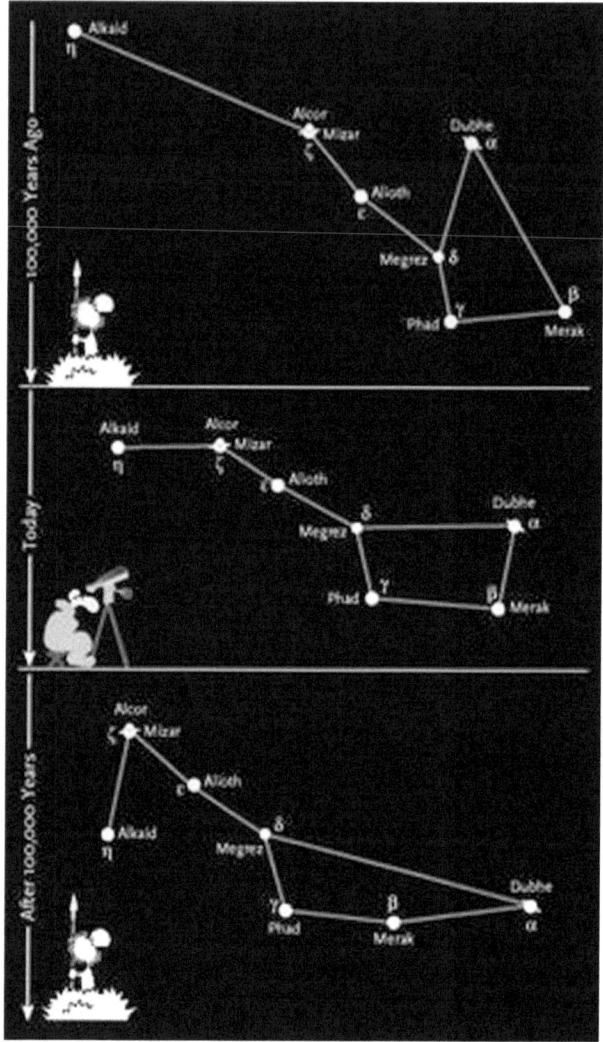

Fig. 3.3 The shape of the Big Dipper 100,000 years ago, today, and 100,000 years in the future (http://www.springerimages.com/Images/Physics/1-10.1007_978-0-387-85355-0_2-4)

along with Alioth, Alcor and Mizar in the handle all belong to the Ursa Major Moving Group. "I find that in parts of the heavens the stars exhibit a well-marked tendency to drift in a definite direction," wrote Richard Proctor, who discovered the phenomenon. It extends through Ursa Major and far beyond into constellations as disparate as Pisces, Cepheus and Triangulum Australe.

Why do they do this? Because the Moving Group is what remains of a lump of stars formed in the same area at around the same time, a process which we will explore in much more detail in a later chapter. We will even be able to observe a minute part of it at work. Most stars in the sky move alone – ours does – but the forces of isolation have yet to separate the Ursa Major siblings. It is a young group, around one tenth of the age of our own Sun, and it is moving towards the galactic center at some speed.

It would be fascinating to find out if there were other features that bound these stars, a family identity, as it were. It happens that an odd pattern of abundances has been noted – they seem to have more of the element barium in their spectra than might be expected, but less copper. This might relate to the conditions in which they were first formed. We will look at our model of star formation in another chapter.

The Rest of the Bear

There has been a lot of science for just a few observations so far. It's bound to be that way as we acquire the tools with which to understand the complex entities that we call stars. Now, though, we can return to the Big Dipper asterism and observe the bowl stars, the four that in Arabic nomenclature make up the body of a funeral bier. Delta Ursae Majoris, the star Megrez, is the dimmest of the quartet while the star below it, Phecda or Gamma Ursae Majoris, is the most alike to our Sun. If you imagine a line between Megrez and Alioth and follow it in the magnification provided by a finderscope, you will come to the star 71 Ursae Majoris, often listed by its catalog reference as HIP60584. It shines a beautiful yellow-gold under medium magnification.

The other side of the asterism offers two stellar treats. In one corner stands Dubhe, Alpha Ursae Majoris, a star some 30 times the size of our Sun and 300 times more luminous. It is not part of the Ursa Major Moving Group and has a companion orbiting it at about 23 AU. (The Astronomical Unit, or AU, is a measurement of space. The distance between Earth and the Sun is taken as 1 AU.) This star is classified as a red giant, spectral class K, a type of star we will return to later. Beneath Dubhe lies Merak, Beta Ursae Majoris. A 1998 study of this star confirmed that it has a dust cloud in orbit around it, a possible precursor to the formation of planets. Together, Dubhe and Merak are sometimes called the Pointer Stars, as a line between them extends north to Polaris, the Pole Star.

The two stars Megraz and Phecda on the handle side of the bowl, meanwhile, point northwards straight at the star Alpha Draconis in the constellation Draco the Dragon. Northern-based amateurs should be able to spot it easily and, although a somewhat inconspicuous object in itself, it is interesting for astronomical history because this star was once the Pole Star. How so? As Earth slowly shifts its axis of rotation, a process called axial precession, the star nearest to the North Pole slowly changes. Alpha Draconis marked this spot, albeit some time ago, in about 30,000 BCE. Polaris came to replace it, and in turn Polaris will be supplanted in some thousands of years by the star Gamma Cephei in the constellation of the whale.

Another line from the asterism stars, this time between Dubhe and Merak, points southwards towards the naked eye star 47 Ursae Majoris. It is roughly half way between the asterism and a bright star on the same line, Regulus in the constellation

Fig. 3.4 M81 and M82 imaged through a TOA150 and Hutech modified Canon 40D, three 20 min exposures at 200ASA (Credit: Robert C Price,www.robertprice10.com)

Leo. Take a few minutes to find 47 Ursae Majoris because far beyond the reach of our sight there are three planets in orbit around it, identified in data assembled at the Lick Observatory. The hunt for such exoplanets is one of the most dynamic areas of contemporary astronomy, and they will form the subject of a dedicated chapter later in the book. For now, it is quite something just to know that they are there.

Ursa Major reveals even deeper secrets. Return to the familiar asterism and imagine a diagonal line bisecting the bowl between Gamma and Alpha Ursae Majoris. Follow this line above the bowl and through a telescope you will be able to make out a pair of galaxies, M81 and M82 – M in this context refers to the list of deep sky objects created by Charles Messier (1730–1817), the Messier Catalog. M81, also known as Bode's Nebula after its discoverer Johann Elert Bode (1747–1826), is a bright spiral galaxy observable as a hazy disk with a bright center. M82, informally the Cigar Galaxy, is mottled in appearance, much longer than it is tall, with dark lanes cutting through areas of brightness. The two deep sky objects are close enough to be affecting each other, a feature of galaxies that we will spend much more time observing in later chapters.

Why a Bear?

The Big Dipper is obvious, but the Great Bear far from so. Yet the Dipper is a mere asterism and the Bear a constellation, one of the 88 that parcel out the sky.

The boundaries of the constellations were fixed in 1922 by act of the International Astronomical Union. They had to resolve some overlaps whereby one star had found itself in two constellations, but for the northern hemisphere at least – the shape of the southern sky emerges out of a different story, which we will tell later – they had a long history upon which to draw. This history reaches back in written records to the ancient Greek astronomer Eudoxus of Cnidus (approximately 410–355 BCE), whose account of the stars is preserved by the poet Aratus (approximately 310–240 BCE). Aratus was perhaps not an observer himself, but he lists the constellations of his day in his *Phaenomena* along with hints as to how to move from one to the next. "Let the left shoulder of Andromeda be thy guide to the northern Fish," is the sort of advice he offers to those who want to find the constellation Pisces, "for it is very near."

Neither Eudoxus nor even Aratus invented all the constellations. They were drawing upon traditions already ancient in their day. But just how ancient is a matter of conjecture. The 12 constellations of the zodiac and those more southerly constellations visible above the equator derive probably from Mesopotamia and the priest-astronomers of Babylon and Assyria from about 1200 BCE. Other constellations, such as those gathered around the legend of Perseus, may well belong to the Greece of Eudoxus's day.

Ursa Major carries a Greek story. Aratus tells one; other Greek mythologies preserve alternatives. But the Bear may well be much older than ancient Greece. It is a strange fact of the culture of the stars that identifying this part of the sky with the Bear unites a broad swathe of peoples from Greeks and Basques to Algonquin, Cherokee and Arctic communities. This is all the more remarkable because, as we've said, the Bear is far from an obvious shape. Scholars have suggested that some sort of original legend may have been transmitted very early in human history, with different stories growing out of some common point. It was perhaps carried from Eurasia to America over the Bering Strait during the last ice age.

A Muddle of Stars and People

The constellations are like that. They preserve fragments of the past, a muddle of some cultures to the exclusion of others. Few people can claim to embody this mix of cultures through which we look at the stars more completely than Donnolo the Doctor.

Shabbetai Donnolo was born in southern Italy in the year 913 in lands controlled by the Byzantine Empire of Constantinople, to a family of strong Jewish identity that had absorbed much of the culture of the Latin world. To add to this amalgam, Donnolo himself was abducted from his town by Moslem raiders in 925 and taken into exile, where he was able to absorb Arabic learning, too. Donnolo was a voracious, even insatiable, reader. This is how he describes his instinct for study: "My heart bade me explore the science of the Greeks, the Arabs, the Babylonians and the Indians. I did not rest until I have copied out books by Greek and Macedonian scholars in their original language."

What did he do with all that he learned? Donnolo was a man of many parts – medical doctor, herbalist and poet, but his heart lay in studies of astronomy and astrology. The great Arabic scholar al-Biruni had established in his *Elements of the Art of Astrology* that nobody could understand stars until they had first mastered where they were. This was so-called 'natural astrology,' which al-Biruni, Donnolo and the whole of the culture of the time distinguished from 'judicial astrology,' the use of the stars to explain or predict events. In two books, Donnolo studied first the constellations and then the whole universe: "The sphere turns the constellations… arranged set to occupy the width of the firmament. The constellations are permanently fixed and attached to it."

We might find some of Donnolo's ideas a little strange. He was convinced that the movements of the stars were controlled by a dragon-like force, the *tli*. He believed in astrology's predictive power, even if we now regard that as unscientific superstition. But he started by looking up to the stars, and gained from his understanding of the constellations and planets something of the nature of humanity:

"Just as God placed the two lights and five planets in the firmament of heaven, so he created in man's head two eyes…. The right eye is like the sun and the left eye is like the moon. The right nostril is like Mercury, the left nostril like Mars. Tongue, mouth and lips are like Jupiter, the right ear is like Venus, the left ear is like Saturn."

We are unlikely to make these same comparisons. But when we talk of Ursa Major and identify the stars, we are no less than Donnolo seeing shapes in the sky that provide meaning and a shared language common to astronomers. It is one drawn out of overlapping cultures, many of them now lost and forgotten. But they are still with us. The Great Bear may not look like a bear, but a bear he will remain.

Chapter 4

The Telescope Revolution

Observation: The Moons of Jupiter
Significance: Galileo's observation reshapes the Solar System
Science: The Galileo mission, Io and Europa

Astronomy is the science of remote things. Except for the fruits of sample-return missions such as *Apollo* and meteorites gathered from the surface of Earth, we cannot touch our data. This makes the telescope not merely useful but essential and therefore powerful. Here is Galileo, writing in 1610: "This new artifice of the spyglass, derived from the most recondite speculations of perspective, brings visible objects close to the eyes."

The telescope was the first physical apparatus that extended human perception beyond natural limits. It has blurred the distinction between near and far to the extent that we now decorate our desktops as easily with distant galaxies as with images of our own world. The telescope is an instrument of revolution.

This chapter is about that revolution and specifically Galileo's role in it. We will repeat in two parts Galileo's observation of the four satellites of Jupiter that as a group are named the Galilean Moons in his honor. By immersing ourselves in what Galileo saw, we can understand how his observations led to profound changes in the way human beings came to understand the heavens and so themselves. From this, we will go on to look at how the same moons have again inspired a revolution in the understanding of the Solar System since exploration of the Jupiter system first by the *Voyager* probes and then by the *Galileo* mission.

M. Marett-Crosby, *Twenty-Five Astronomical Observations That Changed the World: And How To Make Them Yourself*, The Patrick Moore Practical Astronomy Series, DOI 10.1007/978-1-4614-6800-4_4, © Springer Science+Business Media New York 2013

The Jeweled Necklace

There are two straightforward things to check before looking for the moons of Jupiter. First, Jupiter itself needs to be visible above the horizon and, second, the Moon has to be in a phase other than full. You will not need a telescope to make this observation, but if you are using binoculars, see if you can find some way to steady your hands so that you get a good, long look. A tripod is best, but the roof of a car or a patient friend's shoulder will do just as well.

Identifying Jupiter is straightforward. An almanac, astronomical magazine or any one of many online resources will tell you whether Jupiter is visible and where in the night sky it is to be found. It is one of the brightest objects in the night sky. New observers will need to know that Jupiter, like all Solar System bodies, save for our Moon, will appear to the naked eye as a point of light like a star.

Among many others, note the resource provided by Sky & Telescope magazine at http://www.skyandtelescope.com/observing/objects/javascript/jupiter. This provides a map identifying the moons and a list of upcoming occultations and transits.

Fig. 4.1 Jupiter and the four Galilean moons (Credit: Allan Bell)

Look now for the bright starlets close to the disk of the planet. These are the Galilean moons Io, Europa, Ganymede and Callisto. They might be lined up on display to form a glittering necklace – as Galileo expressed it, describing this in his observation of January 7, 1610: "[T]hey intrigued me because they appeared to be arranged exactly along a straight line." On another night, the pattern will be different.

You can establish which moon is which by referring to wavy-line diagrams found in all the resources, which show the movements of the satellites around Jupiter in the center. On any one night, the order of the moons on either side of Jupiter will vary, and one or even two may be concealed by Jupiter's mass.

Io is the innermost moon of Jupiter, 5.9 times the radius of Jupiter from the planet and taking 42.5 h to circle the planet. Its radius is almost exactly that of our Moon and so is its mass density, but the similarities end there. Io, as we shall see, is a rocky world with no craters on its surface. A further 3.58 radii of the planet away lies the next satellite, Europa, whose orbital period is pretty nearly twice that of its inner companion. It is the smallest of the quartet and also far and away the brightest, smothered in deep frozen water ice. Ganymede is the monster of the family and the largest moon in the Solar System. With a radius of 2,631 km, it is broader than the planet Mercury. By contrast, its density is extremely low, 1.942 kg m^{-3} to Europa's 3,014 kg m^{-3} and Io's 3,528 kg m^{-3}, implying that it must contain large quantities of light material, probably liquid water or water ice. It is the only Solar System satellite with its own magnetic field. Beyond Ganymede lies Callisto, orbiting Jupiter at about 1,880,000 km and almost equal in diameter to Mercury. It shows the least evidence of change on its surface.

There are a few things to look for in the movements of these satellites. The first and most striking is when a satellite passes from east to west across the face of the planet. This is a transit, and is often accompanied by a second transit of the moon's shadow, visible as a tiny black dot caused by the angle made between the Sun, the satellite and Jupiter behind. Just as the moons pass in front of the planet, so they also move behind it, disappearing on the west side and reappearing on the east. We will find in a later chapter how important these transits have been in the history of science. Finally, you might notice an eclipse event as a moon enters the shadow cast by Jupiter.

Paths Never Trod Before by the Human Mind

The fact that we have observed easily the scintillating lights around Jupiter, and that we know that they are the planet's four main moons, should not blind us to the fact that none of this was obvious the first time they were seen.

What did it take to find them? First of all, a telescope. Galileo did not invent either the idea or the craft of the telescope, and we will probably never be sure who did. The idea of a telescope had been around for centuries: "From an incredible distance we might see the smallest letters," wrote Robert Record (1510–1558), "so also might we cause the sun, moon and stars to descend in appearance here below." The first definite telescope came into being sometime around September 1608 in The Hague in what is now the Netherlands – an earlier form might have been constructed by Leonard Digges in England half a century before, but the evidence for this is sparse – and it was news of the Dutch manufacture that caught the attention of Italian astronomers. It was a Dutch model that Galileo used to create first his 8X magnification instrument in August 1609 and then the telescope that he describes in his book *Sidereus Nuncius* ("Starry Messenger") of 1610:

> "I progressed so far that I constructed for myself an instrument so excellent that things seen through it appear about a 1,000 times larger and 30 times closer than when observed with the natural faculty only."
> The title of Galileo's book can be translated as "Starry Message" or "Starry Messenger." The evidence suggests that Galileo intended the first but did not at all mind the second, which has become common in English:

Others were looking as well, notably the English astronomer and lunar cartographer Thomas Harriot (c1560–1621). Had Harriot published his discoveries the rivalry between the two men might have been intense.

Galileo's telescope was not easy to use. When Stephen Ringwood reconstructed a telescope such as that Galileo used, he reported "great difficulty finding an object through the eyepiece and keeping it in view. Even the Moon takes a little time to place in the field, let alone planets and stars." Significantly, though, "the four Galilean moons stood out crisply and cleanly." By such effects of optics are revolutions made.

Galileo's gift lay in interpreting what he saw. When around December 1, 1609, he examined the Moon through his telescope, he saw a pattern of light and shade that he understood as mountains and valleys, a rugged landscape. Others who saw the same phenomena a little later could only perceive what they had been taught to perceive, dismissing lunar topography as incidental blemishes on the ever-perfect sphere. Galileo had no doubt, writing that "it is like the face of the Earth itself."

Where after the Moon? Galileo next turned to a question that was to haunt astronomy for a long time – were cloud-like patches of brightness called nebulosity really some sort of cloud or an effect caused by stars shining from very far away? The Milky Way was where that question had to be answered first, and Galileo was able to resolve many, many stars with his telescope. "To whatever region of it you direct your spyglass, an immense number of stars immediately offer themselves to view."

Then, some time before January 7, 1610, Galileo turned his attention to Jupiter. He saw three stars shining alongside the planet.

Which was in itself not very much. They might simply have been more stars. But Galileo did not look just once. Over successive nights, he saw a motion among those stars that did not make sense unless they were connected to the planet. Also, on his fourth observation, he spotted another star. Their motions persuaded him to advance an entirely new idea. He wrote at the end of the month of how he had seen "Four new planets…that move around a very large star, like Venus and Mercury, and perhaps the other known planets, do around the Sun."

Starry Messenger was written to tell this to the world. A short book collated hurriedly, it is an account of the building of the telescope and of these observations. There was not time to do more. Speaking of the constellation of Orion, Galileo promises,

"I will put off this assault until another occasion." A lot of the book is comprised of Galileo's drawings of the Moon, the stars and finally the four moons of Jupiter, the moons shifting position on either side of the planet in exactly the way we observe today.

Galileo disseminated his other discoveries, too. News of how he had seen the planet Venus came in the form of a Latin anagram, *Haec immature a me iam frustra leguntur oy*, which when translated informed perplexed recipients, "I am now bringing these immature things together in vain, oy." His answer to the puzzle was *Cynthia figures aemulatur mater amorem*, "The mother of love emulates the shape of the Moon." Since the 'mother of love' was Venus, Galileo was saying that he had seen phases of Venus through his telescope.

A little before this, Galileo had published the word *Smaismrmilmepo-etaleumibunenuguttaurias*, which Kepler translated as revealing something about Mars. Thomas Harriot made many attempts to work out what was being said without the least success. It turned out that Galileo was announcing that Saturn had protuberances. He had observed the planet's rings.

Galileo became a celebrity. Within a year or so, some of those who could had repeated his observations and agreed with what he had seen, if not with what the observations meant. Galileo used his success to curry favor with the Medici princes of Florence, writing to Cosimo in the preface to *Starry Messenger*, "The Maker of the stars Himself admonished me call these new planets by the illustrious name of your Highness…. Since under your auspices, Most Serene Cosimo, I discovered these stars unknown to all previous astronomers, I decided by the highest right to adorn them with the very August name of your family."

The flattery did not endure. The names by which we know the Galilean moons were those applied to the discoveries by Simon Mayr (1573–1624, sometimes spoken of by his Latinized surname Marius).

Not everyone liked Galileo. "His skull is affected, delirium fills his mind," wrote Martin Horkey in April 1610. "His optic nerves are destroyed because he has scrutinized minutes and seconds around Jupiter with too much curiosity of presumption." Horkey, who made something of a fool of himself by attempting to rebut Galileo in print, attacked not only the man but also the intention and the message.

What was it about the message that caused such a reaction? It was not that Galileo's findings added to the Copernican model of the universe, at least not initially. *Starry Messenger* is pro-Copernican but does not prove the case. Rather, his observations subverted a cherished distinction, one with two millennia of pedigree, between the things of Earth and the things of the heavens.

How did they do this? Galileo was claiming that the Moon looked like Earth, undermining the difference between the two bodies. Jupiter's moons clearly circled their own planet much as the Moon does our own. He was saying there was more to know that could not be found in Aristotle. One man with a telescope could crack open the Solar System and make centuries of natural philosophy look inadequate. It is not hard to understand why nothing was the same again.

Starry Messenger did also provide compelling support, if not final proof, for alternative models of the Solar System to the Ptolemaic spheres. The Jupiter satellites, contentedly revolving around the planet itself, was not neutral in the great debate over how the heavens really worked.

Moving Stars

To observe the orbits of the Galilean moons, the best method is to do exactly as Galileo did and sketch the pattern made by the satellites. It is not that easy to achieve. You will find that the nearest of the moons has an orbital period of a few days, but the more distant takes 2 weeks or more. You can check your results against the resources. Don't be surprised if you are off by quite a bit.

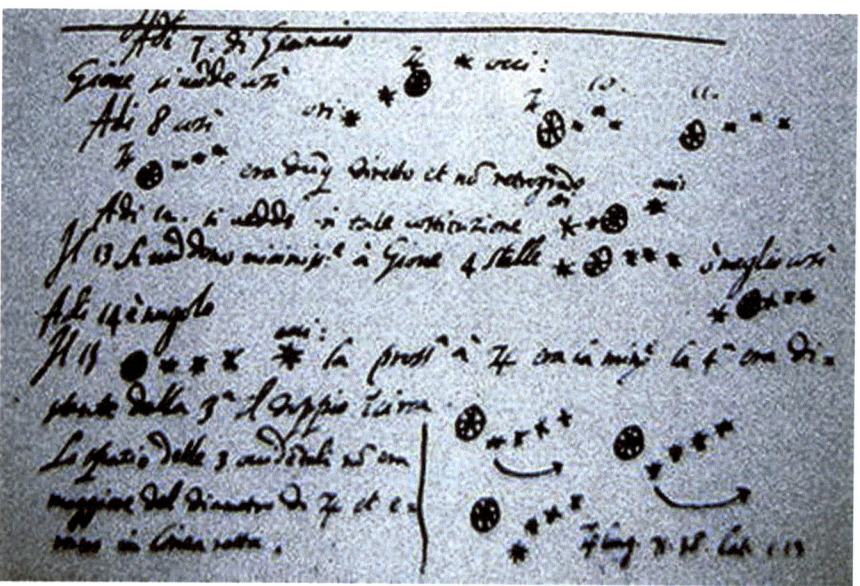

Fig. 4.2 Galileo's sketch of Jupiter and its moons (http://www.springerimages.com/Images/Physics/1-10.1007_978-1-4419-6803-6_1-22)

There are other Jovian moons – 50 are confirmed and named, with 14 or so still listed as provisional. But the non-Galilean satellites are far outside the reach of amateur instruments. The first of them was not spotted for some 280 years after Galileo. The discovery of Amalthea is credited to the American astronomer E. E. Barnard in 1892. A few more were discovered early in the twentieth century, but the planet's smaller moons were not spotted until probes visited the Jupiter system.

Galileo Condemned

"Your actions are observed minutely," a correspondent warned Galileo, "and are being published around the world." What he observed had come to matter very much.

Galileo was not converted to a Copernican view of the Solar System by what he saw through his telescope. He was already a Copernican when he looked. In 1597, Galileo was already telling Kepler, "for many years I have come to the same opinion as Copernicus and from that point of view the causes of many natural effects, which undoubtedly cannot be explained by the common hypothesis, have been revealed to me." The argument made sense of the data more simply than did its rivals. But the sight of the moons of Jupiter was new evidence in favor of the theory. In *Starry Messenger* he says that the observation was: "An excellent and splendid argument for taking away the scruples of those who, while tolerating…the Copernican system, are so disturbed by the attendance of one Moon around the Earth while the two together complete the annual orb around the Sun that they conclude that this constitution of the universe must be overthrown as impossible."

In other words, now that Jupiter was known to have moons, there was nothing so special in Earth having one as well and therefore possessing two centers of rotation, itself for the Moon and the Sun for both of them. But as Galileo moved towards embracing the Copernican model more openly, some of those who were watching began to fear not only for the destruction of the Aristotelian heavens but also for the apparent contradiction between Copernicus and the Bible. In 1616, the Holy Office of Pope Paul V let it be known that the idea of Earth's motion was 'at least erroneous in faith,' and Copernicus' writings were placed on the Index of Forbidden Books.

Galileo, duly warned, kept away from the controversy, but his 1632 book, *A Dialogue Concerning the Two Chief World Systems*, returned him to its heart. He was placed on trial in 1633 in the ecclesiastical court of the popes of Rome, and on June 22 he was condemned as "vehemently suspected of heresy, namely of having held and believed a doctrine which is false and contrary to the divine and Holy Scripture: that the sun is the centre of the world and does not move from east to west, and the earth moves and is not the centre of the world."

He spent the remainder of his life under house arrest.

Did Galileo's telescope prove that Copernicus was right? It did not convert everyone. The Tychonic universe that we described in Chap. 1 also made sense of the observational evidence, and as late as 1674, Robert Hooke was wondering if it did not make better sense than Copernicus' idea of Earth in motion. The final proof of the Copernican universe came only when astronomers were able to measure parallax, the angle of the stars caused by the rotation of Earth. Galileo sought for this but could not find it with the telescopes at his disposal. How it was found is another story for a later time.

Condemnation could not undo observation. But with Mayr's classical names attached to them, the incendiary properties of the Galilean moons seemed to be quenched.

Which did not mean they were forgotten. Their role in the discovery of the speed of light will occupy us in a later chapter. Meanwhile in the eighteenth century attention gathered around the ratio of the moons' orbital periods. For Io, Europa and Ganymede, these ratios were weirdly exact. The first director of the Stockholm Observatory, Pehr Wilhelm Wargentin (1717–1783), established them as 1:2:4, meaning that for every orbit made by Ganymede, Europa made two and Io four. These suspiciously perfect numbers were explained by their gravitational relationship, now known as a Laplace Resonance after Pierre-Simon Laplace (1749–1827).

Human sight returned to the moons via satellites, first NASA's *Pioneer 10* and *11* probes and then on board *Voyagers 1* and *2*. Their flybys in the 1980s proved that the moons were not dead lumps of rock but four worlds quite different from each other and from anywhere else we knew. While *Voyager 2* was heading for the outer Solar System in July 1977, a new Jupiter orbiter and probe was being approved by the U. S. Congress. The probe, renamed *Galileo,* was waiting at the Kennedy Space Center on January 28, 1986, the day the shuttle *Challenger* blew up. It eventually lifted off from Earth on October 18, 1989, on board the shuttle *Atlantis.*

The probe weighed 3 metric tons. Most of that was propellant and a heat shield. It proved its worth long before reaching the Jovian system by probing the lunar south pole and making a close flyby of asteroid 951 Gaspra.

Galileo also detected life on Earth, although not very much of it. At maximum resolution, the probe could not identify any cities or artificial lighting, and the only forms of life directly observed were at the Great Barrier Reef. But its instruments did detect high levels of oxygen and methane in the atmosphere, as well as emissions in the radio spectrum.

Not that its journey was without problems. The high gain antenna, designed to transmit data at 140,000 bits per second, failed to unfold, reducing the communication potential to the secondary system, which operated at 10–20 bits per second. This meant that the science teams had to choose how best to use meager resources, and they opted for detailed images rather than attempting to better *Voyager's* global maps. Given what Galileo achieved, it is staggering to think how much more there might have been.

Arrival day was December 7, 1995. "It's sort of like your birthday," Carl Sagan wrote. "Your parents have permitted you to see the pile of presents but you can't open them yet; you don't know what's in them." Scientists are still unwrapping them now, presents from Jupiter and the moons which, after their slumber in the human mind, would once again overturn a lot of what we thought we knew.

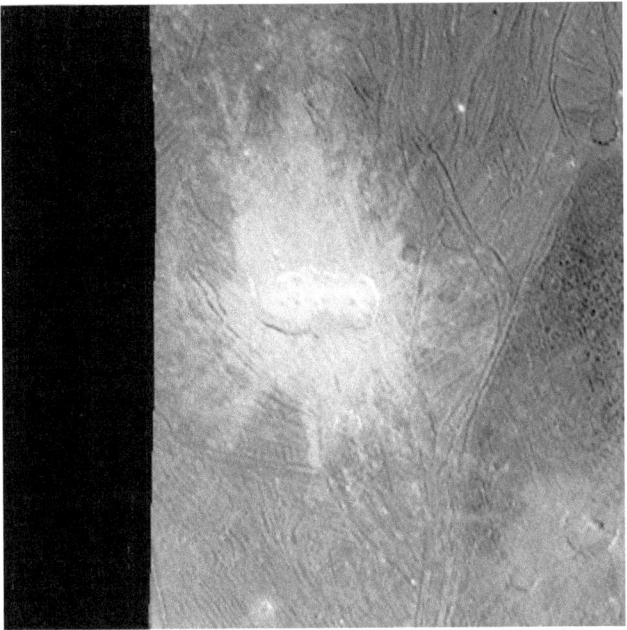

Fig. 4.3 The Nanshe Crater, Ganymede (Credit: NASA)

What did the *Galileo* probe see? We are going to look at the science of two of the four moons in a little more detail, but here are some of the headlines. Io is a volcanic world with active plumes and lava flows. Europa looks like the surface of the Arctic Ocean and may be the only world apart from our own dominated by water. Ganymede is a world of bright and dark lands, with a differentiated interior and the greatest diversity of impact events in the Solar System. Look at the Nanshe crater in the image above, with its seven central pits. Callisto, finally, is the simplest of the four, cratered and ancient but perhaps with its own salty ocean beneath the surface.

The Fires of Io

Jupiter's innermost moon is dominated by volcanoes. When this was discovered by Voyager scientists in 1979, the effect was to end forever the distinction between the geologically alive Earth and the dead elsewhere in the Solar System. Io achieved for twentieth-century science what the moons together did for the seventeenth.

Voyager identified plumes of material emerging from the satellite's interior. The *Galileo* mission established that this moon is the most active volcanic body in the Solar System, putting out some 30 times the energy of Earth.

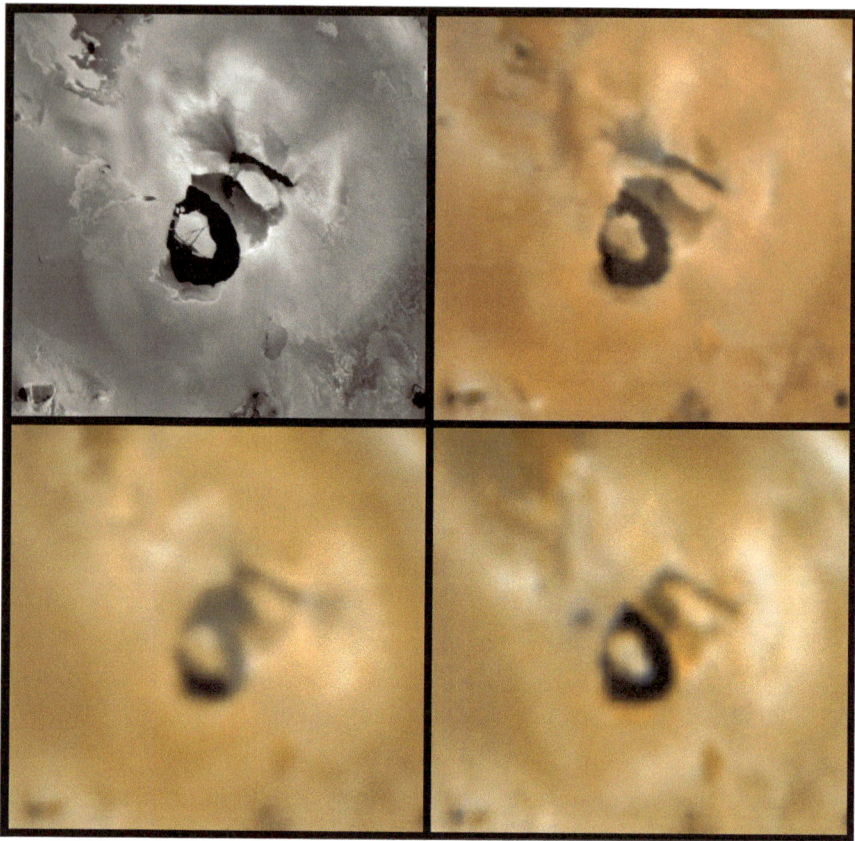

Fig. 4.4 Four views of the volcano Loki Patera on Jupiter's moon Io showing changes seen on June 27th, 1996 by the Galileo spacecraft as compared to views seen by the Voyager spacecraft during the 1979 flybys (Credit: NASA/USGS)

Io's most intense volcanic site is the Loki Patera, probably a giant lava lake releasing somewhere between 10 % and 25 % of the total heat being radiated from the surface at any time. Loki seems to have an island enclosed within its horseshoe shape. *Voyager I* saw plumes rising from one end. *Galileo* spotted dark material that had appeared between the visits of the two spacecraft.

Where does Io's energy come from? All activity within the moon stems from tidal heating, the effect of the satellite's interaction with both Jupiter and its fellow Galileans. As witness to the power this has on the surface, Io does not display craters but rather tall mountains and broad plains smeared with sulphurs and silicates. This gives Io its strange, mottled surface.

Fig. 4.5 Io transits in front of Jupiter (Credit: Eric Walker)

Where Io is multicolored to resemble something like a pizza, Europa is bright white, a world of cracked ice punctured by impacts that have left massive craters, Callanish and Pwyll. The *Galileo* probe's sharp eyes were tuned to exploring the endless patterns of Europa's plains.

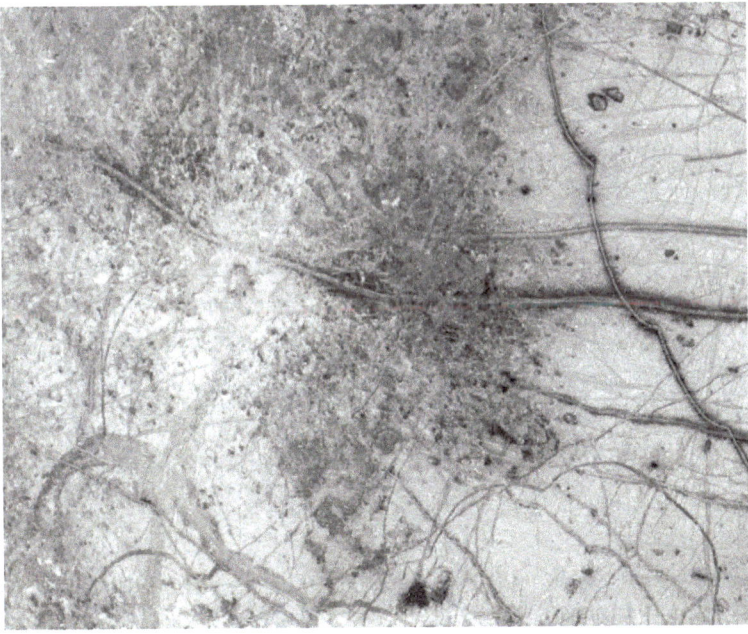

Fig. 4.6 Europa's complex terrain revealed in a Galileo image of 20 February, 1997 (Credit: NASA/JPL)

They look like this. In this image, a new impact crater has impacted with the ice and sent white debris across the greyer surface. It also cuts across the Belus Linea, a 'triple band' characterized by dark outer margins and a bright stripe in the center. The outer margins are diffuse, suggesting that the dark material is a result of geyser-like activity.

Europa's icy crust is thought to conceal a subsurface sea. Energy gathered from tidal interaction may be melting the interior such that the moon harbors vast areas of liquid water, perhaps covering the whole globe.

And where there is water, there may be life. Is there life on Europa? We do not know, but it is no surprise that the European Space Agency has recently announced its intention of following *Galileo* with a new mission, provisionally named JUICE, which will spend 3 years exploring Europa, Ganymede and Callisto. Its questions are big ones: What are the conditions for planet formation? Can life emerge? How does the Solar System work?

Galileo would have been proud.

The Pleasure of Telescopes

There are many moments of excitement in amateur astronomy – the first sight of the Jovian moons, the first comet, the Andromeda Galaxy, completing the Messier challenge. But high in any list must come unpacking and erecting a new telescope. It can lead anywhere, a pathway to the infinite possibilities of space.

The same zeal to discover drove one of the great telescope creators of the nineteenth century, William Parsons (1800–1867), briefly Lord Oxmantown but best known as the Earl of Rosse. He created during the 1840s an instrument known as the Leviathan of Parsonstown, a 72-in. reflector for which he constructed his own alloys. It used mirrors so large he had to build an automatic grinding and polishing machine. In September 1844, a letter to *The Times* reported that the great mirror, some 5 in. thick, was ready, but it was not until early 1845 that it was turned seriously to the skies. "This instrument," an astronomer asked by Rosse to join him wrote, "bids fair to throw a light hitherto not merely unobtainable but unhoped, upon the constitution of the universe."

This was Ireland. It rains there a lot. February of that year provided no clear nights at all. March was better, a 'lucid interval' according to one of those observing with Rosse. Non-astronomical visitors also got in the way. But late in the month or in April, Rosse was able to observe and critically to draw a galaxy we know as M51. He used a smaller telescope to establish the outline and the Leviathan to capture the detail. Rosse saw the connection between the galaxies and spiral structure. A new age had begun in the understanding of far-distant objects, and the Leviathan was to make a contribution, if not a conclusive one, to the same debate over nebulosity to which Galileo had contributed centuries before. It was the Leviathan, also, that enabled Rosse to draw many deep sky objects, achievements described as 'epoch-making' by John Louis Emil Dreyer (1852–1926).

Dreyer knew what he was talking about. The creator of the *New General Catalogue of Nebulae and Clusters of Stars* (the *NGC*), Dreyer looked back in 1909 to exactly 300 years earlier, "one of the most remarkable epochs in the history of astronomy," when Galileo turned his telescope towards the heavens and: "As it were in a twinkling of an eye, the whole aspect of the universe changed."

The moons of Jupiter now seem quite near. An amateur telescope can peer at least as far as Lord Rosse's Leviathan. But the wonder of those starlets forever rearranging themselves around Jupiter should never leave us. They changed how we understand Earth and the heavens, and they might do so again.

Chapter 5

The Hunter's Stars

Observation: The Orion constellation
Significance: Stars have a life cycle from birth to death
Science: Scientific method, the Hertzsprung-Russell diagram

Our fifth observation is going to work a little differently. Before we go outside to set up the telescope, we are going to think about how science works.

It is tempting to think that scientific advance proceeds by a succession of illuminations, various apples falling from various trees and hitting great scientists, mostly men, on their heads so that they cry out 'Eureka' and find that they have solved gravity, parallax, the age of the universe or how to get a spaceship to Pluto. Science becomes a story built out of individual genius. Everyone is wrong until one person is suddenly right.

Sometimes there are eureka moments and singular great individuals. But not very often. Science makes progress like most other areas of human endeavor – by way of muddle.

This muddle is given a grand name – the scientific method. Thinkers propose models that explain particular observations. Then they design experiments that test the theory, experiments capable of being repeated, both to protect against individual false results and also to eradicate effects due to particular circumstances such as location or atmospheric conditions. A theory that fits the data might then challenge other theories, leading to new hypotheses, new experiments and so on. This is less dramatic and more stumbling than apple-and-eureka – it just happens to be the way to get things right.

One of the best definitions of the scientific method comes from the pen of Lewis Carroll, author of *Alice in Wonderland* and *Sylvie and Bruno* (1889). The narrator of this latter book is at times confronted with some very strange things and makes

M. Marett-Crosby, *Twenty-Five Astronomical Observations That Changed the World: And How To Make Them Yourself*, The Patrick Moore Practical Astronomy Series, DOI 10.1007/978-1-4614-6800-4_5, © Springer Science+Business Media New York 2013

sense of them in this way: "First accumulate a mass of Facts: and then construct a Theory. That, I believe, is the true Scientific Method. I sat up, rubbed my eyes, and began to accumulate Facts."

Making sense of the stars worked in just this way. It isn't a smooth story. But one tool that was created out of questions, muddle and doubt has become one of the most powerful in all astronomy. We can use it to deepen our understanding of the stellar observations we are going to make in this chapter among the stars of the constellation Orion.

Making Sense of the Stars

Our observations in Ursa Major demonstrated how stars differ one from the other. This was a decisive change in the way that humans looked into the sky. Where before there had been uniformity, now the stars were varied, their inner lives revealed by spectroscopy. Comparing stellar properties enables astronomers to distinguish between common and unusual stars.

What if stars are so different that they cannot be compared? One of the great debates of nineteenth-century astronomy concerned whether or not stars shared enough properties to be considered as one entity. Work in the 1860s on the light emitted by the stars suggested that a small number of fundamentally different groups of stars existed, so distinct as to be separate creatures. There were also very many stars, too many, possibly, to discover anything that applied to them all.

As we have seen, there were many astronomers at work on creating classification schemes. At Harvard, Annie Jump Cannon took huge steps forward in this, while Antonia Maury identified what seemed to be not one but several 'courses of development' across the histories of different stars. She perceived for the first time that there was evidence of stellar evolution.

Maury's work led to a real but almost secret breakthrough when the Danish astronomer Enjar Hertzsprung (1873–1967) combined her scheme with other statistical studies to produce in 1906 a diagram that compared stellar luminosities with temperature on a graph.

Hertzsprung was very nearly not an astronomer at all. At age 20, he sold all his astronomy books and later wrote: "Nobody imagined that I should ever become an astronomer." Perhaps that was why he published his diagram in a periodical so obscure that no one noticed it. He did write eventually to E. C. Pickering, who replied politely but skeptically although he knew that another astronomer, the American Henry Norris Russell (1877–1957), was working on comparable ideas. Russell's diagram first appeared in 1910; he presented his work to a wider scientific audience in 1913. Out of these separate endeavors was born the Hertzsprung-Russell, or HR, diagram, a way of comparing stars.

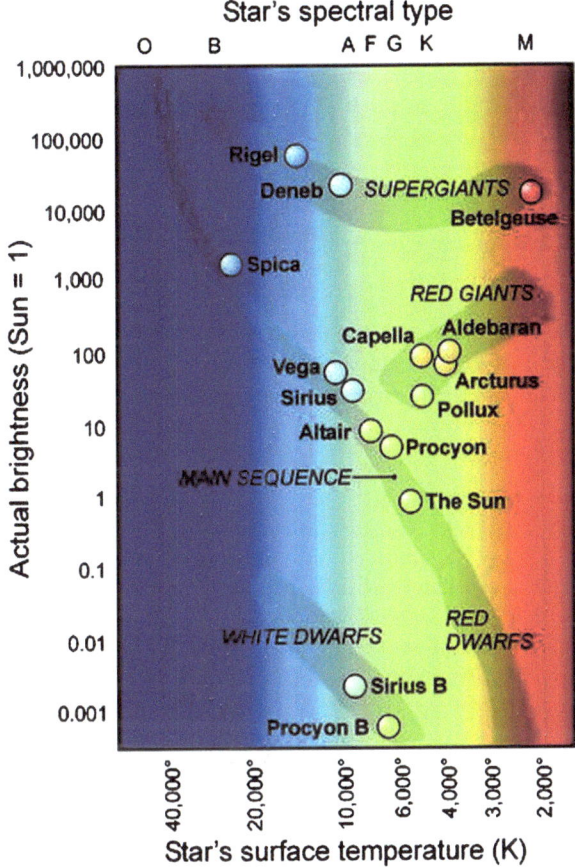

Fig. 5.1 The Hertzsprung-Russell diagram (Credit: Peter Grego http://www.springerimages.com/Images/Physics/1-10.1007_978-1-4419-5592-0_6-16)

Hertzsprung published in the journal *Zeitschrift fur Wissenschaftliche Photographie* (Journal of Scientific Photography). In later years, British astronomer Sir Arthur Eddington wrote to Hertzsprung, "One of the sins of your youth was to publish important results in inaccessible places."

The Hertzsprung-Russell diagram charts two properties. Along the horizontal axis is a measurement of the temperature of stars in Kelvin. The vertical axis tracks luminosity. From these two pieces of information, temperature and brightness, an equation establishes the radius and therefore the size of any star using the Stefan-Boltzmann constant. Note that the temperature line runs backwards, so that the hottest readings are nearest the inner corner. Both scales are logarithmic. The dots on the graph represent where stars are found, and the names of principal stars are marked.

In the diagram above, the quantity of data is not large. But it gives us some important general results. At the heart of the graph is a diagonal line where some 90 %

of stars are found. This is marked on the illustration with the label *Main Sequence*. Above and to the right lies the land of the giants, with distinctions between different types. For example, supergiants are more luminous than red giants of comparable temperatures. Meanwhile, below the Main Sequence line, live the dwarfs. Giants and dwarfs – the language is Russell's.

There are lots of ways in which the HR diagram represents a transformation in our understanding of the stars. It reveals that they clump into groups and can therefore be compared. We can place our own star in the context of others. Above all, the HR diagram allows us to step outside the limitation of our own mortality. An individual star exists for far longer than the span of human history, and there is no way that we can watch one evolve – there isn't time. The HR diagram reveals that stars have a life story comprising youth, middle and old age. The overwhelming majority of stars are in their Main Sequence or middle-aged phase.

The Hunter

Let us now take the HR diagram and use it under the stars.

Almost however you judge the different constellations – clarity of pattern, variety, brightness, interest – Orion wins. It is also easy to identify, forming an obvious pattern that looks like a human figure or perhaps an hourglass. Since it crosses the celestial equator, Orion is visible throughout the world and dominates the winter nights in the northern hemisphere.

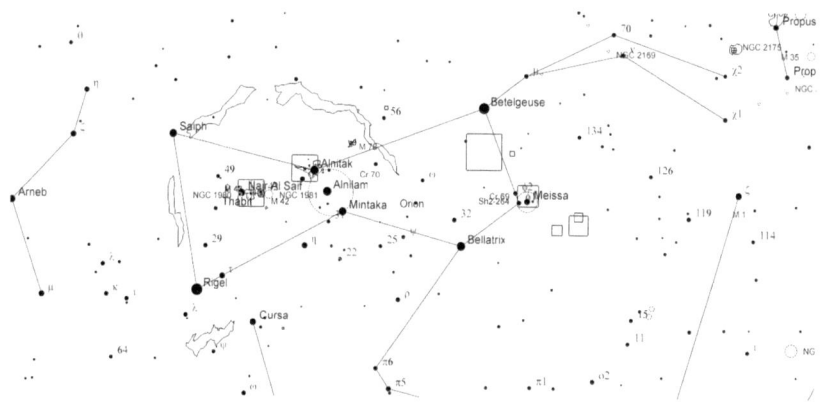

Fig. 5.2 The main part of the Orion constellation (Cartes du Ciel)

A quick view of the constellation through binoculars reveals the variety of stars. The most prominent star is the giant Alpha Orionis, or Betelgeuse, which shines red to the naked eye and splendidly so in any magnification. Its brightness and size both vary for reasons we will explore later in this chapter. Alongside Betelgeuse and forming the other 'shoulder' of Orion is Gamma Orionis, the star Bellatrix, whose blue-white light makes a beautiful contrast with its neighbor. Bellatrix's place on the HR diagram reveals that it, too, is a giant star, though of a different type. Between and slightly above a line made by these two stars is the group of stars marking Orion's head.

The most distinctive feature of the constellation is formed by the belt, a line abreast of three white stars. From the uppermost star beneath Bellatrix these are, in order, Delta, Epsilon and Zeta Orionis, bearing the names Mintaka, Alnilam and Alnitak. Sigma Orionis lies below Zeta, marking Orion's sword and the location of the Orion Nebula, which will concern us in another chapter. The hourglass asterism then widens to the two stars at the foot, of which the brightest, shining at magnitude +0.1, is Rigel. Rigel is a supergiant, one of the most massive stars we know. Its companion is Kappa Orionis, the star Kaph. Stretching out from Bellatrix are the stars that form the hunter's bow, and above Betelgeuse lie a web of stars that are identified as his raised arm.

Familiarity with Orion is a good starting-point for other observations in the northern sky. A line drawn between the two brightest stars, Rigel at the bottom of the constellation and red Betelgeuse at the top, extends beyond Betelgeuse towards Castor and Pollux, the twin stars at the head of Gemini. The three belt stars point downwards at the brightest star in the whole night sky, Sirius in Canis Major, and in the other direction they guide the eye towards first Aldebaran in Taurus and, beyond this star, to the blue-white cluster of the Pleiades, or Seven Sisters. Another line, this time from Bellatrix through Betelgeuse, leads towards Procyon, the alpha star of Orion's lesser canine companion Canis Minor. Finally, a line from the outstretched foot Saiph on Betelgeuse's side of the constellation through Betelgeuse and beyond points at Capella, the alpha star of Auriga.

Fig. 5.3 A sunlike image of Betelgeuse, made with NASA's Hubble Space Telescope, taken with the Faint Object Camera on March 3, 1995 (Credit: Andrea Dupree (Harvard-Smithsonian CfA), Ronald Gilliland (STScI), NASA and ESA)

The image above takes us intimately close to Betelgeuse, the eighth brightest star in the night sky and one of the most studied. Its place on the HR diagram reveals that it lives at the extreme of stellar possibilities, a red supergiant varying in magnitude between +0.2 and +1.2 as the star expands and contracts. Betelgeuse belongs to the M spectral type, a group embracing a vast range of stars from supergiants like this one to dwarfs. They have different futures – M supergiants are all losing mass and face violent destruction, while red dwarf stars have lifetimes longer than the galaxy. These will, in effect, never die. What unites all M stars is that they are cool, below 3,800 K.

Variously spelled and pronounced in at least three ways according to the *Oxford English Dictionary,* the name *Betelgeuse* is a corrupt form of an Arabic original that meant "the hand of the female one." Because the constellation became male in Greek myth, the name was transliterated to mean something like "the armpit of Orion."

How big is Betelgeuse? The mass of a star is difficult to determine, and estimates have varied. But a recent study suggests that Betelgeuse is some 11.6 times the mass of our Sun, with a margin of error of +5 and −3.9. This massive star is cool, busy shedding vast quantities of material into space, as revealed by another image of the star, this time taken by the European Space Agency's VLT (Very Large Telescope) array.

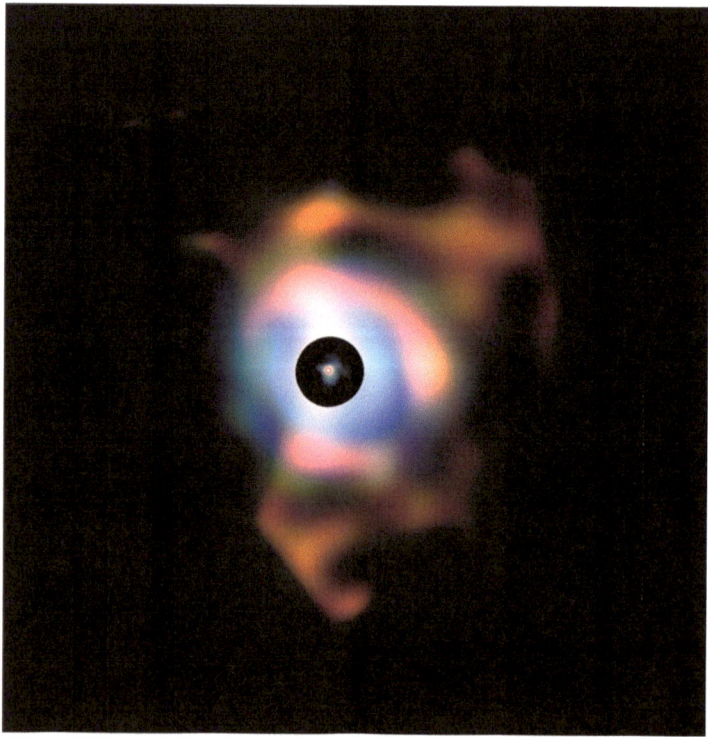

Fig. 5.4 The nebula around the bright red supergiant star Betelgeuse, created from images taken with the VISIR infrared camera on ESO's Very Large Telescope (VLT) (Credit: ESO/P. Kervella)

The star itself has been occluded here so as to reveal the chaos existing around Betelgeuse in ultraviolet as plumes are cast out of the star's interior. These are Betelgeuse's writhings as it approaches its death. Sometime in the next million years or so – that's pretty soon in astronomical terms – Betelgeuse will explode into a supernova. It will shine more brightly than the Moon and be visible in our daylight. But only for a time. Then Betelgeuse will grow dark and disappear.

In the meantime, it would be interesting to know how far away Betelgeuse is. But these measurements are hugely difficult, as we will see. Data gathered from the VLA and the *Hipparcos* satellite both suggest that Betelgeuse is 643 light years from Earth, with a margin of error either way of 146 light years.

Betelgeuse's companion shoulder star is Gamma Orionis or Bellatrix, a Latin name meaning "female warrior." Bellatrix is also a giant star but inhabits the other end of the spectral scale. It is a B-type star, a group characterized by vastly high temperatures as can be observed from this star's place on the HR diagram. Bellatrix may have a surface temperature of 22,000 K. With a magnitude of +1.64, few pairs exhibit the contrasting states of stars more than these.

The belt stars also make a compelling trio. The stars at either end are good targets to try to split into multiple components, as both of them are binary stars. The lives of binary stars will be a topic we will look at in more detail later. In really good observing conditions, the belt star beneath Betelgeuse, Zeta Orionis or Alnitak, has a ghost in attendance on it, the Flame Nebula, NGC 2024. The nebulosity extends all around the star but is particularly striking above it. The trick to observing the Flame is to keep the star itself outside your telescope's field of view; otherwise its brightness will flood everything else. The east side of the nebula is the more defined, with dark intrusions that Alnitak does not seem to block. The Flame is part of a great treasury of nebulas around Orion – we will spend a good deal of time exploring these in another chapter.

It is well worth finding the star Sigma Orionis just below Alnitak. There are some lovely stars around it and three seemingly very close, a great focus for smaller telescopes. Sigma can be elusive, but it is worth some perseverance. It makes a fine end to an evening with this constellation.

Don't ignore Orion's feet. Rigel, Beta Orionis, is a blue supergiant some 773 light years away from us. It has at least three companion stars, but as Rigel A is 500 times brighter than the next, they are a real challenge to separate.

See if you can find the Hunter's head. Swing back to Betelgeuse and Bellatrix and find the star above them that seems to form an isosceles triangle. This is Lambda Orionis and has two close attendants well worth searching out. They form part of a cluster of some 20 stars called Collinder 69, of which six also stand out through a small telescope.

Before we finally leave the Hunter for his prey here is a final challenge. If you focus your telescope on the three belt stars and then move up as if towards Betelgeuse, you will find, at the top end of your field of view, a star that never received a name or Bayer classification and so is known as either HR1988, HD38529 or HIP27253. It's not frightfully exciting in itself, a mere subgiant, but it is significant in that two planets have been observed in its orbit. There is no chance at all of directly observing exoplanets with amateur equipment, but it is still special to see a star with a confirmed planetary system.

Orion's Prey

Hunters hunt, no less in the sky than in the woods and forests, and Orion's prey constellations are worth exploring as we expand our grasp upon the many types of stars.

One target of his labors is the constellation Lepus the Hare. Lepus is not an easy constellation to envision, but the bright star Alpha Leporis lies directly beneath Alnilam in the center of Orion's belt and forms a tidy triangle with the two feet stars of the hunter. Another huge star, some 14 times the Sun's mass, the star shines yellow-white, the color characteristic of the F class, to which also belong well-known stars such as Polaris and Procyon. F-type stars are a transition type leading towards the A type, of which Sirius is the finest example.

Just above Alpha Leporis and directly beneath Rigel is the star Mu Leporis, with a magnitude of +3.5 and, like Bellatrix, a B-type star. It lies 186 light years from the Sun and is interesting for its odd chemistry, exhibiting strong lines of mercury and manganese in its spectroscopy. Mercury is present in abundances some 100,000 times that of our Sun in stars like this, which have the label Chemically Peculiar, or CP. In part caused by gravitational settling of some elements and the rising through the stellar atmosphere of others, CP stars such as Mu Leporis demonstrate how the composition of stars can vary across the galaxy, giving the lie to the appearance of a lot of uniform white dots in the high sky.

If you find Alpha and then Mu Leporis and continue along the line they make, you will be able to identify a distinctively red star, R Leporis. This is Hind's Crimson Star, "a drop of blood on a black field" according to Burnham's *Celestial Handbook*. It's tricky to capture in the telescope. You might find it easier to follow an alternative route by drawing a line from Mintaka through Rigel. It is worth persevering to see this star, both for its color and for the science that causes it. Hind's Crimson Star returns us to Betelgeuse's M spectral type and introduces us to stars that vary in magnitude. R Leporis moves between magnitudes of +6 and +11 over a period of 432 days, at its reddest when it is least bright. It is a carbon star, at a stage in its evolution when carbon is being exhaled into its atmosphere. This point of red in the deep night is a sign of how stars evolve.

John Russell Hind (1823–1895) was an English astronomer who observed both variable stars and asteroids. On September 13,1850, he spotted a new asteroid and named it after Queen Victoria, an act of patriotism that broke the convention of not naming stellar objects after the living. But Victoria stuck, W. C. Bond of Harvard coming to the Queen's aid by pointing out that Hind might have been referring to the mythical goddess of victory. He wasn't, but the excuse worked.

Returning to Alpha Leporis, almost directly below is the bright (well, bright for this region) Beta Leporis. If you draw another line between these two stars, you'll see that it points down towards the globular cluster M79.

There are different views on how visible or not M79 actually is. A higher power on an average telescope gives a good sense of M79's structure, and something really powerful (16 in. or so) can open out the cluster's magnificent core. It is concentrated and bright, a challenge to split.

Fig. 5.5 M79, taken with a Ritchey-Chretien 32″ telescope and SBIG STL-11000 CCD camera (Credit: Jim Misti, www.mistisoftware.com)

The cluster occupies an interesting place on the HR diagram. It is full of very blue stars burning helium in their cores. They are not far from their collapse. There is also recent evidence of the presence in M79 of a type of star called a PAGB, a red supergiant on its way to becoming a white dwarf.

The science of the stars is rich and complex. Observing Orion and Lepus emphasizes the variety of stars in the night sky. The HR diagram and the spectral types enable us to place stars with others like them, identifying common traits and dissimilarities. These observations beg the question of where stars come from and how they were formed. This will be the main theme of our next chapter.

It will not have escaped your notice that star-naming is a complicated business. It encompasses the poetry of names such as Betelgeuse and Bellatrix to functional but overlapping alphanumeric designations. How did we get into this state?

When we name the stars, we are making use of different layers of human history. Some history, at least – star names from cultures such as the Chinese, Japanese, African or Native American scarcely appear at all. Some attempt to repair this cultural exclusivity has been made with planetary features, as we will see.

About 70 % of the star names in modern use are Arabic. About 19 % are Greek or Latin. This reflects the importance of Arabic star maps in the history of astronomy. Some of the names have specific meanings. They might refer to vanished patterns of constellations once seen in the sky. Other names have carried astrological significance.

Names have appeared and disappeared with the waxing and waning of particular maps. Most people can name the twins of Gemini, Castor and Pollux. Few recall their alternate names Ankelar and Abrachaleus. Some catalogs preserve their Greek names; in that culture, they were Apollinis and Herculis.

It was the printing of star charts and figures of constellations that started to promote a more ordered approach. Perhaps the most significant mapmaker of the skies was Johann Bayer (1572–1625), whose star atlas *Uranometria* of 1603 introduced the system of double labeling discussed in Chap. 3. But Bayer's system covered only the stars he could see, and as telescopes grew more powerful, new labels were needed. Some stars have numbers only, and these are the Flamsteed numbers after the eighteenth-century observer John Flamsteed (1646–1719), whose atlas *Coelestis Britannicae* was published posthumously in 1729.

We have already seen how a code system is used to reference the nebulae listed in the Messier Catalog. These are commonly referred to by the M1 to M110 designations. Messier's list is augmented by the *New General Catalogue (NGC)* of J. L. E. Dreyer and by the Collinder Catalog of open clusters published by the Swedish astronomer Per Collinder in 1931.

For stars, a number of new catalogs offered new systems for identifying the formerly unseen. The HD system stands for Henry Draper, HR For Harvard Revised, which is a bright star catalog, then BD for Bonner Durchmusterung, GC for General Catalog and so on. Southern stars were not included in the BD system, and so are identified by CD (Cordoba Durchmusterung) and CPD (Cape Photographic Durchmusterung) numbers. Meanwhile Betelgeuse is variously Alpha Orionis, HR2061 and BD71055.

Confusing? Yes, but these are the stars as we have made them, and perhaps as they should remain. We have looked and we have named and we have tried to understand. None of these works will ever be completed.

Chapter 6

Home and Next Door

Observation: The Milky Way and the Andromeda Galaxy
Significance: The limits of our own galaxy, the existence of others
Science: Galaxies, the great debate, basic shapes of galaxies

We have discovered our place in the universe. *Voyager 1* and *2* have reached points so far from the Sun that they are able to detect the spectral line of atomic hydrogen. It is something we have detected in other galaxies but never before in our own because of the smothering effect of the solar wind. That measurement is a sign of our belonging to the Milky Way.

Two galaxies are the concern of this chapter. Observing and understanding them are among the most important discoveries that human beings have made. They show us what our home looks like, how far it stretches and where it stops. They reveal the shape and science of galaxies distant from our own.

The observations come first. They are accompanied by visits to other targets and constellations. After this, we will revisit the 'Great Debate' that helped to establish the size of the Milky Way and the existence of separate galaxies. Then we will come to the science of galaxies today.

Walking the Milky Way

Almost everything that we see at night, all the stars that are not other galaxies or other deep sky objects, belongs to the Milky Way, so in one sense it is almost all we ever see. But the specific observation is that of a stretched ribbon of light across the sky, the band of our home galaxy.

M. Marett-Crosby, *Twenty-Five Astronomical Observations That Changed the World: And How To Make Them Yourself*, The Patrick Moore Practical Astronomy Series, DOI 10.1007/978-1-4614-6800-4_6, © Springer Science+Business Media New York 2013

The name "Milky Way" is a translation of the classical Latin *Via Lactea*. The Greek name was Milky Circle, and it was as *Milky Circle* that the name is first recorded in English. Other languages preserve different memories of seeing. In Chinese, our galaxy is the "Silver River" and in Cherokee "The Way The Dog Ran Away," with the white trail of stars representing cornmeal stolen and then dropped by the fleeing hound.

The most important precondition for seeing the Milky Way is a dark sky. Our cities, with all their flashing signs and bright lights to tell us where to go, stop us from seeing where we are. This is Galileo, writing in his *Starry Messenger:* "The Milky Way...with the aid of the spyglass, may be observed so well that all the disputes that for so many generations have vexed philosophers are destroyed by visible certainty, and we are liberated from wordy arguments....To whatever region of it you direct your spyglass, an immense number of stars immediately offer themselves to view."

Fig. 6.1 The Milky Way, Meade LX50 8″ SCT telescope with f3.3 focal reducer (Credit: Eric Walker)

This image captures just one part of the Milky Way's spread. The picture shows a stretch of the galaxy near the constellation Cygnus the Swan, where the *Kepler* spacecraft is currently searching for planets around other stars. Cygnus and other constellations prominent in the northern summer are good places to observe the galaxy's sweep, running above the square of Pegasus and towards Sagittarius. Between Cygnus and Scutum it is possible to observe a dark lane carved through the light – this is the Milky Way's great rift, where interstellar clouds block our view. Any optical aid splits the milky haze into individual stars, as Galileo found, but there is always more light beyond and around them. It's never possible to see to the end of the galaxy.

What are we looking at? It is hard to envision the Milky Way because we are in it. The observation has been compared to trying to see the shape of a mountain from a tunnel carved deep within it. Indeed the band of the Milky Way is something of an illusion, because it gives us the impression, reassuring but false, that Earth sits at its center. In fact we lie above the central plane of the galaxy's disk and are looking down into the disk from one of the galaxy's spiral arms.

Sombrero Galaxy • M104

Hubble Heritage

NASA and The Hubble Heritage Team (STScI/AURA) • Hubble Space Telescope ACS • STScI-PRC03-28

Fig. 6.2 M104, the Sombrero Galaxy (Credit: NASA and The Hubble Heritage Team (STScI/AURA))

This magnificent image taken by NASA's Hubble Space Telescope might help to locate our observation of the galactic river. This is not the Milky Way but a galaxy known as M104, or more informally, the Sombrero. It lies in the constellation Virgo but is far beyond our galaxy. The Sombrero displays itself to us side-on, revealing a clear band of dust at its rim and then a structure, the plane of the galaxy rising to the central bulge or core. These are the characteristic features of a spiral galaxy, of which the Milky Way is one. The river of light that we observe is the galaxy's plane. Its bulge lies in the direction of Sagittarius.

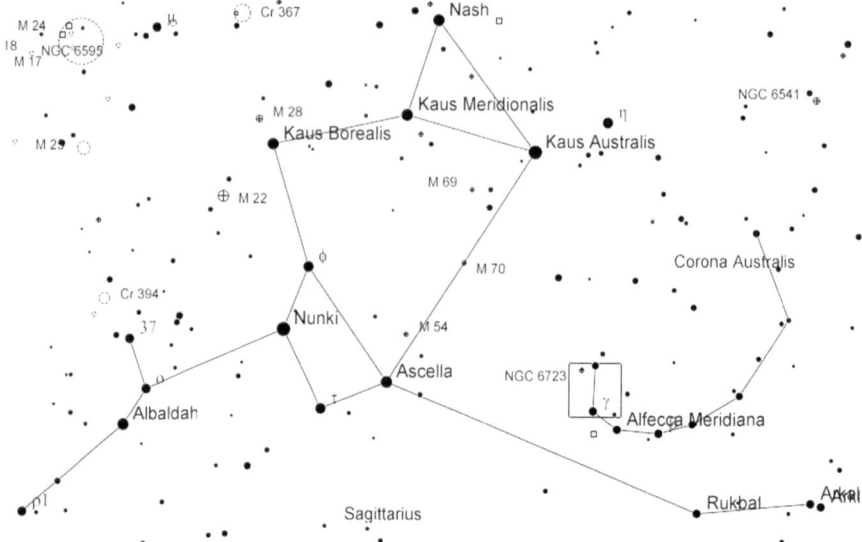

Fig. 6.3   The main part of the constellation Sagittarius (Cartes du Ciel)

Sagittarius is where the most spectacular sights of the Milky Way are to be found. It also contains more Messier objects than any other constellation. It has the largest southern declination of any of the zodiacs, but if visible above the horizon it is not too hard to navigate. Its central stars form an asterism remarkably akin to a teapot, with the Milky Way like steam emerging from its spout. Another way of exploring the constellation is by way of its upturned Big Dipper asterism – the Great Bear stars turned upside down. Four Sagittarius stars make the bowl and two the handle.

Sagittarius also contains the black hole thought to lie at the center of the Milky Way. We will return to this in a later chapter.

The stars of Sagittarius' bow are worth exploring. Its northern tip is marked by K-type Lambda Sagittarii, the star Kaus Borealis, which is at a late stage in its stellar evolution, fusing helium into carbon and oxygen in its core. As an observation, Kaus Borealis is a golden treat for the eyes. A line drawn between Mu and Lambda Sagittarii scrapes past two globular clusters close to Lambda, M28 and M22. The latter was the first globular cluster to be discovered and is a naked-eye object, just. A telescope resolves into separate stars, though it's only possible to observe a fraction of the 83,000 gathered here. Using a small instrument to observe M22 is to follow in the footsteps of Sir Edmund Halley, who in 1716 described the cluster as "small but very luminous."

The phrase 'globular cluster' describes a tight ball of bound stars. We will observe and explore globular clusters at length in a later chapter.

Mu Sagittarii is also the guide to a great Messier object, the Lagoon Nebula M8. This lies a little to the southwest of the star and makes a triangle with Lambda. M8 is a fine binocular object, dominated by the O-type star 9 Sagittarii shining some 8,000 times more brightly than the Sun. Through good binoculars or a small telescope, it is possible to make out dark lines among the nebular clouds and lots of shapes, including a striking hourglass feature. The Lagoon Nebula is a rich observing object and a great target for amateur astrophotography, as this image demonstrates.

Fig. 6.4 M8, a 20 min exposure taken using a Takahashi FSQ 106 telescope at f/5 and an SBIG ST8-XE CCD camera (Credit: Paul and Liz Downing, www.paulandliz.org)

Returning to the line of stars making up the bow, pause at Delta Sagittarii, the giant K-type star Kaus Medii, which has a beautiful deep yellow, almost bronze color. It contrasts with Epsilon Sagittarii, Kaus Australis, a chemically strange star with a deep blue appearance.

The glow of the Milky Way suffuses most of Sagittarius until it ends precipitously in the darkness of the Great Rift. The galactic center is, from our perspective, a little nondescript, lying towards the western edge of Sagittarius where the constellation merges into Scorpio, but there is a window through which we can peer towards it, the Small Sagittarius Star Cloud M24. M24 is a naked-eye observation in all but the most polluted skies. It is a patch of haze lying in the north of the constellation, above the teapot asterism along a line formed by the stars Epsilon (Kaus Australis) and Delta Sagittarii. It is not really an object but a serendipitous absence of inter-vening dust, a hole that enables us to look past both our own and the Sagittarius-Carina arms of the Milky Way towards the central bulge. An amazing number of stars become visible, and there are many distinct objects to observe among them, most notably the dark nebulae classified as Barnard 92 and 93 towards its north. There are patterns and swirls in the stars to make this an enduring and fascinating sight. It is a sword of sight to pierce the galaxy's heart.

Observing Andromeda

It is easier to see another galaxy than our own. The Andromeda Galaxy M31 lies 2.5 million light years from Earth and is the most distant object observable in northern skies without a telescope. It has been known about and watched for many centuries. Indeed, this galaxy is more apparent than the constellation in which it lies. Andromeda is an indistinct grouping of stars, but it lies between two more obvious star formations, the square of Pegasus and the W-shape of Cassiopeia.

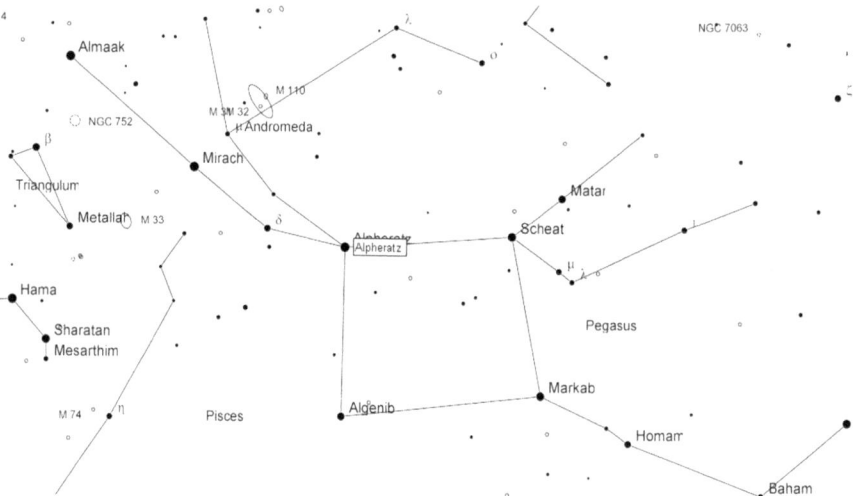

Fig. 6.5 The Constellation Andromeda (Cartes du Ciel)

To reach M31, it is easiest to start at the square of Pegasus, an obvious four-cornered asterism. Despite its name, the upper left star of the square is not in Pegasus at all. This is Alpheratz, the alpha star of Andromeda.

Alpheratz is an interesting star in its own right. It consists of two stars, far too close to be separated optically, that are orbiting each other in just 96.7 days. Perhaps because of their proximity or for other reasons, the brighter star of the pair exhibits an atmosphere rich in manganese and mercury. Alpheratz is also a useful star, for it lies at the beginning of a clear chain of three, all of which appear to the naked eye to be about equal brightness and form a nearly straight line that Bayer identified as successively Alpha, Beta and Gamma Andromedae, with little Delta lying between and just below the line binding Alpha to Beta. Becoming familiar with this 'string of pearls' asterism is the key to exploring the constellation.

To identify M31, find the middle star of the three bright ones in the chain. This is Beta Andromedae, the star Mirach. Move up from Mirach towards Cassiopeia's W and the galaxy is nearly impossible to miss. With the naked eye, it is possible under reasonably dark sky conditions to see its smudge of nebulosity. This glimpse of Andromeda is tantalizing, for all the surrounding stars are part of the Milky Way. Mirach, for example, lies 350 light years away, but Andromeda is millions of light years beyond the Milky Way's rim, a vast galaxy that rewards patient observation through any and all magnifications that you have. You should see a bright, egg-shaped center and the bright bulge of the galactic core. Stretching out to either side of this runs a crayon-drawn band of light that often exceeds the width of the eyepiece under higher powers.

There is a lot of structure visible in the galaxy, especially in the outer regions. This structure has been observed in great detail, and there are useful web resources that can guide you through the bright patches and dark lanes.

To explore Andromeda in more detail, consult the atlas prepared by Paul Hodge at http://ned.ipac.caltech.edu/level5/ANDROMEDA_Atlas/Chart5.html, but also note http://www.regulusastro.com/regulus/papers/m31/, where further images and charts are available.

Fig. 6.6 M31 and M32, Vixen ED81S f/5,2 on Takahashi Em-200, Canon 350D (Credit: Tom Victor Kolkin, http://home.broadpark.no/~tomvk/)

M31 is not alone. Binoculars will reveal the glimmering companion galaxy M32 below but close by its greater neighbor. M32 has a habit of fuzziness, but a larger aperture telescope will open out some structure. It also provides scale, for M32 is about 20,000 light years from M31. Another galaxy, sometimes called M110 although Messier never included it in his original list, lies above M31, a little further away than M32. NGC205 is a very faint binocular object but good for a telescope.

Before leaving the sky to understand more of galaxies, return to the line of three stars in Andromeda to make a further observation. The third star in the line of pearls stretching from Pegasus is Almach, Gamma Andromedae, a fine double star to separate under modest magnification. The brighter star shines yellow-orange, sometimes gold, at about magnitude 2, while the second is a brilliant blue-white. It was the sight of these contrasting stars that led William Herschel to write: "The striking difference in the color of the two stars suggests the idea of a sun and a planet, to which the contrast of their unequal size contributes not a little."

Hershel was not to know that the visible pair constitute only half the story. Almach is a quadruple system some 200 light years from Earth.

A Battle for the Stars

The stars between upper Andromeda and the constellations of Cassiopeia and Cepheus were once the site of dispute between ambitious European powers. In 1679, Augustin Royer created a grandiose constellation here called Sceptrum et Manus Iustitiae, the Sceptre and Hand of Justice, as a tribute to the Sun-King Louis XIV of France. 115 years later, this same patch of stars was rendered in Germany as the constellation Frederici Honores, the Glory of Frederick, to honor the Prussian King Frederick the Great. Neither of these constellation names survived, but it is curious that the relics of both monarchs ended up in this corner of the sky.

Permanent amidst the politics is NGC 7662, the aptly named Blue Snowball Nebula, which through small telescope looks much like a star but under higher magnification reveals an attractive, blue-green glow. It lies about 3,000 light years from Earth and is a magnet for deep-sky photography. The Blue Snowball is best found via Iota Andromedae, which at magnitude +4.3 is one of the brighter stars west of M31. A little further to the southwest from this and a reasonable telescope combined with some patience will reveal the faint spiral galaxy NGC7640. It is edge on to us in the manner of Andromeda and the Sombrero.

Fig. 6.7 NGC7640, LX200 with a 41 min exposure (Credit: John Ambrose, www.johnsastrop ics.com)

How Big Is Home?

As we look at Andromeda through our 'scopes, it seems obvious that it lies outside our own galaxy. It is, of course, not obvious at all. The conclusions that the Milky Way is one galaxy among many and that it is possible to observe objects beyond the Milky Way's limits emerge out of a great scientific debate of the twentieth century. Its outcome changed the way we understood our place in the universe.

Before the advent of twentieth-century technologies, the evidence for other galaxies was far from clear. The Nebular Hypothesis explained observable patches of brightness and the spiral forms as being made of gas rather than stars, gas lying on the edge of the single, vast Milky Way. Some observers even claimed they could detect the spin of the gaseous arms.

New kinds of astronomy weakened this argument. Analyses of Andromeda's spectroscopy did not fit with the gas model, while new and better photographs of spirals indicated the presence there of exploding stars or supernovae. Working with the Mount Wilson reflector, George Ellery Hale (1868–1938) amassed not only evidence but also immensely able assistants, among them Howard Shapley (1885–1972), who specialized in studying the variations in light produced by a certain kind of star, Cepheid variables. We will spend time in their company in a later chapter. For now, it is how they were used that came to matter. Shapley, building on the work of Henrietta Swan Leavitt (1868–1921), photographed star clusters with Cepheids and was able to use them to make measurements of stellar distances.

Shapley's data provoked a storm. It indicated that these clusters lay across a huge sphere and that the center of the sphere lay in Sagittarius. "The Solar System can no longer maintain a central position," Shapley declared, concluding, in a sense, the argument inaugurated by Copernicus. Shapley's measurements had the effect of inflating the Milky Way. He argued that its diameter was as much as 300,000 light years. With this figure, it seemed clear to him that everything lay within the Milky Way's embrace. He argued for a 'big galaxy.'

Others disagreed. The debate on the size of the Milky Way will always be associated with Shapley and another American astronomer, Heber Doust Curtis (1872–1942) of the Lick Observatory. They crossed swords at the April 1920 meeting of the National Academy of Sciences in the Smithsonian Museum, Washington D. C. Each was given just 45 min to make his case, but Curtis wanted "a good friendly scrap" even in this short time.

Curtis had been photographing nebulae for a decade or more, and his evidence implied that the Milky Way was altogether smaller than Shapley had estimated. While he did not deny the possibility that the Milky Way was a far greater galaxy than others such as Andromeda, he used a phrase of Immanuel Kant's and spoke of other, 'island universes.' It implied a limit to our own galaxy. "Debate went off fine in Washington," Curtis wrote afterwards, "and I have been assured that I came out considerably in front."

What was the outcome of this 'Great Debate'? There was no doubt that the Milky Way had a center and humanity was not it. But the limits of the Milky Way were not established until one of Shapley's colleagues, Edwin Hubble (1889–1953), studied hundreds of photographs of Andromeda and found there "dense swarms of images in no way differing from those of ordinary stars." He found Cepheids there and was able to measure their distances sufficiently well to establish that Andromeda had to lay beyond even Shapley's inflated Milky Way. We will return to the significance of this moment in a later chapter, but the point was now proven – the Milky Way was one among many. This many stretched out to the limits of our imagination.

The Shapes of Galaxies

Hubble's achievement was not confined to resolving the Great Debate. He is credited with a system of classifying galaxies that, with some modification, still endures in modern astronomy. Hubble's system, often presented in the shape of a tuning fork, distinguishes between four major groups of galaxies classed by their shape –elliptical, spiral, lenticular and irregular.

Helpful as it is, the tuning fork diagram can give the impression of an evolution-ary chain. So can some of the language around it – Hubble talked of *early* and *late* galaxies. But it is wrong to think of this as expressing an age sequence. Hubble himself knew it was not. "The nomenclature," he wrote in 1927, "refers to position in the sequence, and temporal connotations are made at one's peril."

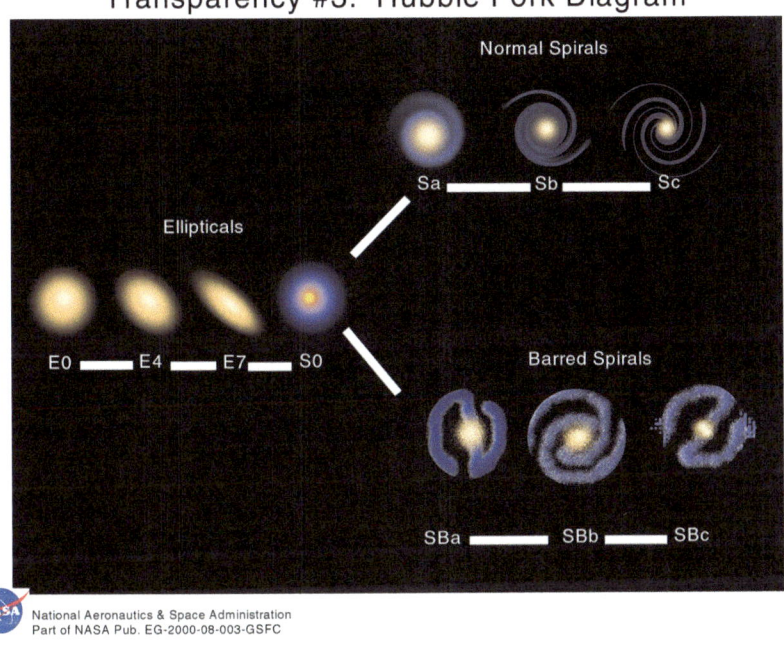

Fig. 6.8 Hubble's fork (Credit: NASA)

Elliptical galaxies, those shaped roughly like an ellipse, form the largest group. They tend to appear featureless, although observations reveal a concentrated light source at the center with brightness tapering towards the rim. They range from the nearly circular E0 type to the most elongated E7. An example of this type of galaxy is M89 in the Virgo galaxy cluster.

More dramatic in appearance but much less numerous are the spiral galaxies. Spiral galaxies contain a circular disc and central bulge, and the disc contains spiral arms. Various subdivisions exist within this group – the now familiar Andromeda Galaxy is a clear example of a spiral.

Lenticular galaxies are lens-shaped with a bulge but no spiral arms. They are often seen as occupying an intermediate zone between elliptical and spiral, although not in terms of evolution. There's no evidence of one galaxy type becoming another.

The final class of galaxies is explained by its name. The irregulars are, well, irregular. There are also more than a few galaxies that have proved unclassifiable, and others, such as the celebrated M87, can be variously classed according to the duration of the photographic exposure. On a long exposure, M87 appears to be elliptical in a fairly ordinary way, but shorter image times reveal a 'jet' emerging from its core. M87 is now classed as a 'peculiar galaxy,' and it will be a deep-sky observation to which we will return.

The science of galaxies puts us in a position to understand ourselves. The Milky Way's classification is difficult because we cannot see it from the outside, but it is labeled as a barred spiral. As for us, we're not the center of it but are perched on the Orion Arm, although whether it is a true arm or a spur is not clear. As to the size of the Milky Way, it depends on what is measured. Modern theory now proposes that the most massive component of our and every other galaxy is dark matter, thought to be in a halo that holds together all the visible components. Since this dark matter cannot be observed, it may be better to ask after the size of the Milky Way's disk. It is thought that this disk is about 100,000 light years in diameter and about 1,000 light years thick, containing somewhere in the order of 200 billion stars.

Is this scale wonderful or terrifying? The human imagination has veered between hope in the vastness of space and fear of what it might contain. Hope usually involves us going out to them as bringers of technology, civilization and law. Fear works in reverse, exploring what we would do if or when they came to us. In 1610, reflecting on Galileo's discoveries, Kepler asked: "If there are globes in the heaven similar to our Earth, do we vie with them over who occupies the better portion of the universe?"

At some later points in this book, we will meet several attempts to answer this very big question.

Chapter 7

Our Red Neighbor

Observation: The planet and orbit of Mars
Significance: The shape of the Solar System, Mars and imagination, the search for life
Science: Martian orbit and surface, the *Viking* experiments

Mars has always mattered more than any other of the planets. We have spread across its surface a mat of hopes and fears, pouring resources into the effort to understand it – more spacecraft, more studies, more everything than anywhere else. Why?

In part, because it is a fascinating observation. Its red color draws us in. Its details appear familiar, with polar caps that wax and wane like ours. William Hershel told the Royal Society in 1784 that, "the analogy between Mars and Earth is perhaps by far the greatest in the Solar System." We have been drawing pictures of Mars since the 1600s. Modern research has added a great deal of detail, but then new questions come up after every set of answers it has achieved. Mars feels inexhaustible. We can never say of it, "I know all about that place."

Mars has also been the focus for the search for life. "If here," the question seems to go, "then why not there?" Where the scientists have not been sure, the dreamers have been certain. We all have an idea what a Martian should look like. Mars has become a chance to start again. Will we make of it a paradise or hell?

This chapter will make two observations of the planet, one straightforward and the second more detailed. In making the first, we will repeat an observation that defined Earth's place in the Solar System. After looking at the planet in more detail, we will follow the thread of the search for life through the modern science of Mars and the visions it has inspired.

M. Marett-Crosby, *Twenty-Five Astronomical Observations That Changed the World:*   75
And How To Make Them Yourself, The Patrick Moore Practical Astronomy Series,
DOI 10.1007/978-1-4614-6800-4_7, © Springer Science+Business Media New York 2013

Does Mars Move Backwards?

Mars is not always visible from Earth, but when it is, the Red Planet displays its distinctive color under any dark sky. Any astronomy magazine or website will tell you where to look. Mars is to be found somewhere near the ecliptic, a line passing through the zodiacal constellations that all the planets in the Solar System follow more or less closely, giving rise to some beautiful alignments in the night sky.

For a good overview of Mars, see Mars Previewer, which can be downloaded via the *Sky and Telescope* website at http://www.skyandtelescope.com/resources/soft ware/3304921.html?page=2&c=y. A useful observation record form is at http://www.gaherty.ca/rogers/Marsobservingform.pdf while http://ralphaeschliman.com, which offers some beautiful maps.

Once you've found Mars, notice its progress through the sky measured with reference to background stars. Magazines and online resources will, once again, indicate the constellations through which the planet is passing. Given time, you will observe something that perplexed humanity for thousands of years but that, once understood, changed fundamentally the way that humans came to understand their place in the universe. Don't tell anyone, but Mars seems to move backwards.

This observation informed the Ptolemaic model of how Solar System worked. To the apparently obvious facts that Earth was stable – we did not seem to fall off – and at the center of the sky, the ancient system added three details to explain the awkward orbits of Mars and the other planets. Firstly, the theory placed Earth slightly away from the center of their orbits, a point called the deferent. Secondly, the model proposed that Mars and each other planet moved around a loop of its own, an epicycle, the center of which moved in a wider circle around Earth. Finally, the motion of epicycle was believed to be uniform with respect to another point, this time called the equant, the same distance again as the deferent was from Earth.

This Ptolemaic system was not ridiculous. As explained first by Georg Peurbach (1423–1461) and then his great pupil Johannes Müller von Königsberg (1436–1476), always known by the Latin name Regiomontanus, the Ptolemaic system offered a sophisticated explanation of what observers actually saw. But it was complicated, and there was always disagreement as to its details. These became especially fractious in the early sixteenth century, when rival systems were created and older ones revived to try to explain new observational data. Many of these models, created by such men as Girolamo Fracastoro and Giovanni Battista Amico, have been long forgotten – Fracastoro did get a crater on the Moon, and we will observe it in a later chapter – but one theory did not get lost. In the end, it triumphed.

Regiomontanus completed first a new translation of Ptolemy started by his master and then penned an *Epitome of the Almagest* in 1463. He had ambitious plans for the renewal of mathematical and observational astronomy, barely started by the time of his death in 1476. It was Regiomontanus' account of Ptolemy that Copernicus both used and praised.

Which was not obvious at the time. When Nicholas Copernicus published first his *Commentariolus* before 1514 and then his *De Revolutionibus* in 1543, he did not shake the world to its foundations. But it proposed a different way to see the skies: "To be sure Claudius Ptolemy of Alexandria, who far excels the rest by his wonderful skill and industry, brought this entire art almost to perfection with the help of observations.... Nevertheless very many things, as we perceive, do not agree with the conclusions which ought to follow from his system."

So Copernicus opened his work, nothing if not respectful of the older systems. In the dedication at the front *De Revolutionibus* – he wrote offering his work to Pope Paul III – Copernicus made much of what the Ptolemaic system did achieve. But:

> Those who know that the consensus of many centuries has sanctioned the conception that the earth remains at rest in the middle of the heaven as its centre would, I reflected, regard it as an insane pronouncement if I made the opposite assertion that Earth moves.

That Earth moves. Copernicus had to be careful. Earth's centrality was as much about theology as science. So against the 'insane pronouncement' Earth might be in motion, Copernicus advanced an alternative history, citing an ancient Greek philosopher Hicetas (ca. 400 BCE–ca. 335 BCE) and the Roman authority Plutarch. Copernicus was telling the Pope that his idea was not a new one, for he knew there would be voices raised against his ideas. "Perhaps there will be babblers who claim to be judges of astronomy although completely ignorant of the subject."

There were also other astronomers working on the same problem. From the great Uraniborg, Tycho Brahe was seeking to establish by observation the evidence for his own or any other of the systems that explained the cosmos. Mars would be the proof, he was sure.

It was a matter of measurement. Tycho worked out that if the planet at opposition, when closest to Earth, was at a distance of about 0.4 AU, then Copernicus' model was the right one. If the planet was further away, that demonstrated Ptolemy's or his own was correct. Tycho was the great observational astronomer of his age, perhaps of all time, and in 1582 he studied and measured Mars carefully.

The result? Mars' position proved that Copernicus was wrong. "By the most frequent and precise observation," Tycho wrote in 1583, "the parallax of Mars was much smaller than required by Copernicus." There – problem solved.

Or not. Tycho repeated the observation during the opposition of 1587. On the morning of March11, Mars was just north of the ecliptic in Leo. But this time, Tycho's measurements confounded his earlier conclusion. As he himself wrote, "I found that Mars displays a greater parallax than the Sun and is therefore closer to the Earth....which is in agreement with the Copernican numbers." Another measurement in 1589 gave the same result.

Tycho was never again able to observe Mars at opposition. The planet did not convert the Prince of Astronomers to the Copernican world view. By 1587, Tycho was advancing his own system, and the Mars observation added weight to its case. Yet Tycho's observations mark an important moment in astronomy, when observation assumed pride of place in the search for understanding. This may seem odd – where else would astronomers look but to the skies? But Copernicus, as noted earlier, built his case more upon the evidence of the past than upon observations, relying upon Ptolemy to prove his case. As Kepler said of him, Copernicus was "representing Ptolemy, not nature."

In fact, the angles Tycho thought he was measuring were well beyond the limits of his instruments. He was not determining parallax at all. No matter, the point was made – it would be observational evidence that would determine the shape of the universe and, therefore, humanity's place within it.

Mars is named after the Latin god of war. The Greeks identified it with their war god Ares, to whose chariot was yoked the horses Phobos (Fear) and Deimos (Terror), the two Martian moons. The Babylonians named it after Nergal, their god of the underworld, while to the ancient Chinese, it was Ying-huo, the planet of fire.

Mars does not move backwards. The regressive drift are a result of the movements of both Earth and the planets. The Sun is also in motion around the galactic center. There is no still point from which to stand and look.

Mars Through a Telescope

Although alluring to the naked eye and brilliantly red under dark sky conditions, it is difficult to get really clear images of Mars. Decent binoculars will enhance the color to a deep red-orange but will not reveal the disk, and Mars' two moons Phobos and Deimos are far too small to be observed.

Fig. 7.1 Mars and the Beehive Cluster (M44), with an Canon 20 Da, 60 s exposure (Credit: John Ambrose, www.johnsastropics.com)

The image above, captured through an 8-in. Schmidt-Cassegrain telescope, demonstrates the possibilities for observing detail on the planet. The point of closest approach between Mars and Earth occurs a week before opposition, and good observations of the planet are available both before and after opposition for several months. Unfortunately, Mars' closest approach every 780 days or so is not always as close as it might be – the planet has the second most eccentric orbit in the Solar System after Mercury, and this transfers into a distance from Earth at opposition that varies by some 25 million miles. Its opposition magnitude varies between −1.5 and −2.9.

To achieve contrast and see more features, it is worth exploring different planetary filters, for example an orange-red filter (Wratten numbers 25 and 29) to enhance the equatorial areas and blue (for example Wratten 38 and 38a, though the ideal filter will vary) for the polar caps.

Mars also repays at least three varieties of patience. The planet is small and quite close by, and if you want to use high magnification, then hang on for good seeing conditions and, above all, steady air. If the stars are twinkling merrily as the air eddies and swirls, then the surface of Mars will appear watery and vague. Secondly, wait a little while for your telescope to get properly cold. This gives the best opportunity to capture the planet's detail. Finally, be patient in this sense – come back to Mars on successive nights. The planet rotates in just over 24½h, which has the serendipitous effect of enabling observers to see almost the same face of the planet at the same time each night. A sketch made on night one grows in detail over the next three or four nights as surface smudges become more familiar.

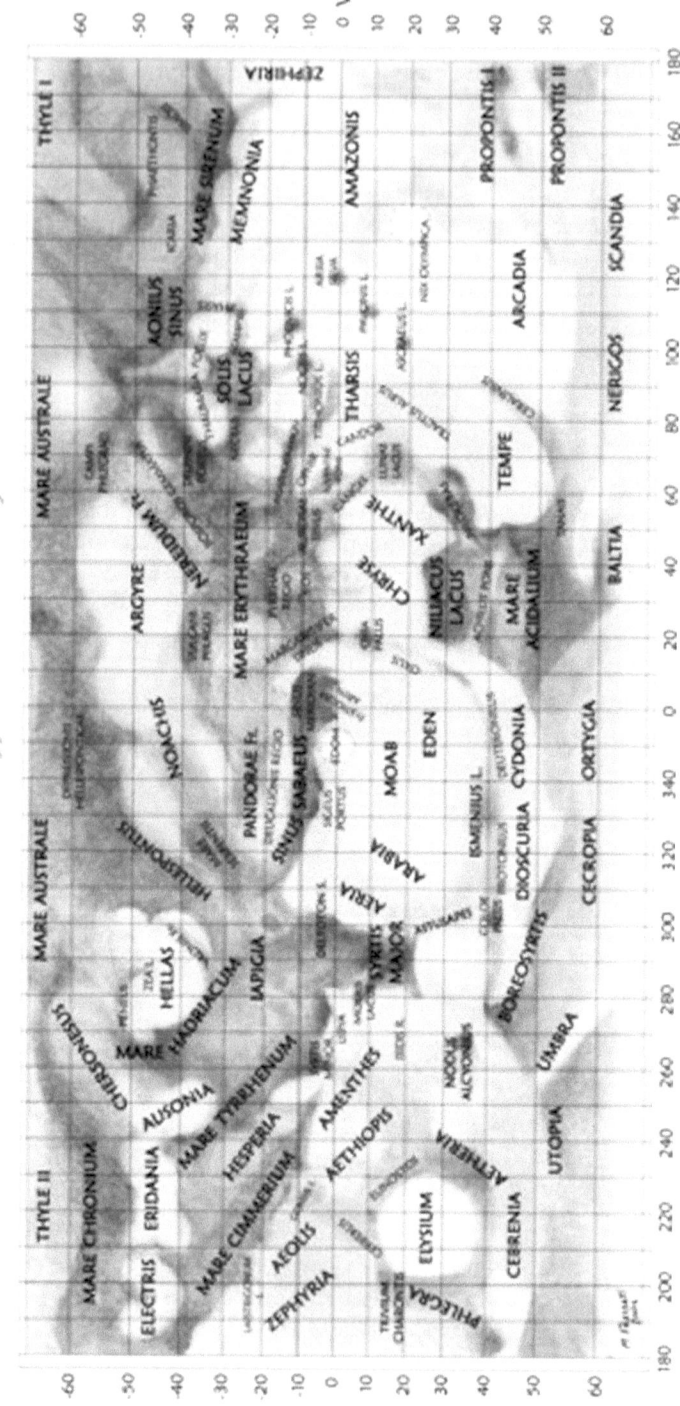

Fig. 7.2 An albedo map of Mars based on the observations made by Italian amateur astronomers from 1988 to 1999 (http://www.springerimages.com/Images/Physics/1-10.1007_978-0-387-76508-2_1-16)

What are we seeing? Because Mars has no significant atmosphere, a 6-in. or smaller telescope will reveal some of the planet's larger surface features. "I pronounce the general configuration of the planet to be very irregular," wrote the Mars observer Eugène Michel Antoniadi (1870–1944), "and shaded with markings of every degree of darkness." The most obvious equatorial mark is Syrtis Major, a dark mark between two lighter areas Arabia and Isidis. The bright desert patch of Hellas lies underneath, and there will be other smudges of darker areas if the seeing is really good. One of the polar caps will always be tilted toward Earth, and this, especially if enhanced by a filter, enhances the sense of Mars' Earth-like seasons.

With a 6-in. scope and careful use of blue and green filters, it is possible to glimpse cloud formations on Mars. These are produced by seasonal heating, water-ice clouds caused by wind passing over the peaks of mountains and even morning fog. Dust storms can smother parts or even all of the Martian surface, and that is something no clever filter can overcome.

The existence of two satellites around Mars was predicted in fiction, first by Jonathan Swift in that part of *Gulliver's Voyages* (1727) when the hero visits Laputa and meets astronomers so able that they can see the two moons. This may have been inspired by mathematics. If Earth has one Moon and Jupiter at that time four, then it was reasonable to imagine that Mars would have either two or three, ensuring a harmonious sequence of numbers. They went with two.

Phobos and Deimos were first observed during the 1877 opposition by the American astronomer Asaph Hall (1829–1907) using the 26-in. refractor at the Naval Observatory in Washington, then the largest refractor in the world. Both are small and reflect very little light, so little was known about them before the *Mariner 7* flyby in 1969. This and subsequent missions have revealed two tiny worlds strikingly irregular in shape and with evidence of heavy cratering – the huge Stickney crater is the most prominent feature on Phobos and is named after Asaph Hall's wife Angeline Stickney Hall. Deimos is most probably a captured asteroid.

It is not only amateur astronomers that struggle with this pair. In the 1980s the then USSR launched two probes to survey and send landers onto Phobos – neither made it to the target, although the second spacecraft, *Phobos-2,* did get close enough to take some photographs. In 2011, Russia dispatched *Phobos Grunt,* an ambitious sample-return mission that would have explored the satellite's surface, including a series of strange grooves, and probed the reason for its low density. But the curse of the two moons persisted. *Phobos Grunt* never escaped its parking orbit above the Earth.

Seeing Red?

On July 18, 2011, people living in the Moroccan desert near Tissint heard two sonic booms and saw a brilliant fireball arcing across the sky. These were the signs of a meteorite striking Earth. The largest lump to survive weighed over a kilogram.

The Tissint meteorite came from Mars. Landing by good fortune in one of the driest places on Earth, it was collected quickly, reducing to a minimum the chances of contamination. "Arguably this is the most important meteorite to have fallen in 100 years," said Dr. Caroline Smith of London's Natural History Museum, where the largest piece of the meteorite is now being subjected to study in a basement archive of other stony visitors from space.

The Tissint meteorite is not red. Beneath the scorched outer layer, Mars turns out to be grey-black basalt. The red we see is caused by a layer of fine dust bright with an iron-rich mineral called hematite.

Is the hematite red because it is rusty? Rust requires both water and an oxidizing agent, and so red hematite might be evidence for the presence, either now or in the past, of those two vital prerequisites for life on the planet. Alternatively, hematite might have been ground down into its present form by the actions of wind storms on Mars. Experiments at the Mars Simulation Laboratory at Aarhus University in Denmark, where the conditions of the planet are recreated as exactly as possible, suggest that Mars' redness could have been caused without either water or any oxygen.

Why does this matter? Because the science of Mars is largely about the search for life.

This search starts with the similarities between Mars and our planet: they have roughly the same amount of land surface area, similar tilts and therefore seasons and both have polar caps. The northern ice cap on Mars is mostly frozen water, while the southern is probably a mix of water and carbon dioxide. Mars currently has no magnetic field, but there is evidence that it once did. So, fueled by the images of fiction, successive missions to the planet have sought for life. They have not found it, but they are still looking.

Areophobia

There was little doubt among early observers that Mars was inhabited, but that these dwellers might be hostile towards Earth occurred to few people before H. G. Wells (1866–1946). None of his books has had a more profound impact than *The War of the Worlds* (1898). The book has been adopted many times by both big screen and small, given an extraordinary life in music by Jeff Wayne and adapted for the radio by Orson Welles to such effect that it caused panic among those who heard it. It is often listed as one of the great books of the twentieth century. This is how the book opens:

> No one would have believed in the last years of the nineteenth century that this world was being watched keenly and closely by intelligences greater than man's and yet as mortal as his own; that as men busied themselves about their various concerns they were scrutinized and studied, perhaps almost as narrowly as a man with a microscope might scrutinize the transient creatures that swarm and multiply in a drop of water. With infinite complacency men went to and fro over this globe about their little affairs, serene in their assurance of their empire over matter.

These were the Martians, "intellects vast and cool and unsympathetic, [who] regarded this earth with envious eyes, and slowly and surely drew their plans against us." When they came, they were both advanced and absolutely alien: "The Gorgon groups of tentacles, the tumultuous breathing of the lungs in a strange atmosphere, the evident heaviness and painfulness of movement due to the greater gravitational energy of the earth – above all, the extraordinary intensity of the immense eyes – were at once vital, intense, inhuman, crippled and monstrous."

The opening of the novel was grounded in contemporary science. An editorial in *Nature* published in August 1894 reported astronomical observations of a light on the planet's surface, which could either be an aurora or a signal from the Martians. Mars was near opposition, so "a better time for signaling could scarcely be chosen."

It is no coincidence that Wells' fictional vision of Mars is contemporary with Percival Lowell's famous book *Mars* of 1895. It contained his speculation that some of the dark markings on the planet might be canals. The story of Lowell's misunderstanding of Schiaparelli's 1877 observations of *canali* has been often told and is not going to be repeated here. It belongs, though, to the same vision of the planet as inspired H. G. Wells.

Mariners and Vikings

When NASA's *Mariner 4* mission set off from Earth on November 18, 1964, its main objective was to overcome the limitations of observing the Red Planet and send back photographs of the Martian surface. It changed our vision of the Solar System.

How? Despite the excitement over the fact that its main objective had been achieved, *Mariner 4's* images were disappointing. It sent back 22 pictures that revealed no life, no vegetation and no Lowellian canals but rather a Moon-like world, cratered and ancient, apparently untouched by geological processes.

Subsequent *Mariner* missions seemed to confirm this. The spacecraft were extremely successful, sending back increasingly sharp pictures of the surface and topographical details of areas such as the Valles Marineris and Hellas Basin, but the overriding impression of Mars remained the same. It seemed that all the visions had been wrong.

Fig. 7.3 Mars, landscape and explorers. NASA's Mars Reconnaissance Orbiter took this image
of the Bonneville crater on January 29, 2012. The lander platform for the Spirit rover is visible
in the *bottom left* corner (Credit: NASA/JPL-Caltech/Univ. of Arizona)

The search for extraterrestrial life on Mars did not end, though. The old dreams
had staying power. Instead it downsized – no more of Wells' creations, but hopes
instead for much smaller forms living in subsurface regions. It was to seek out this
possibility that the Viking missions were conceived. Two landers achieved the
difficult technical feat of touchdown in July and September 1976.

Among many Mars missions launched from the United States, Russia, Europe
and Japan, these are the principal contributors to our understanding of the planet
(all are NASA missions unless indicated): Mariners (1964–1971), Vikings (1976),
Mars Global Surveyor (1996), Mars Pathfinder (1996), Mars Odyssey (2001), Mars
Express (ESA, 2003), the Spirit and Opportunity rovers (2003), and the Mars
Reconnaissance Orbiter (2005).

Like the Mariners, the Vikings were extraordinarily successful. They sent images
from the level of the planet's surface, shrinking the geography so that we could see
Mars from head height. They were automated laboratories, able to sample the Martian
surface in a series of experiments. Both landers carried three main and one subsidiary
element – a Pyrotic Release Experiment, which incubated Martian soil in a simulated
atmosphere and then subjected it to simulated sunlight to see if organic material might
consume labeled carbon, a Gas Exchange Experiment that submerged soil in a
'chicken soup' mix and then tested for emissions, the Labeled Release Experiment,
which added water to a soil sample and monitored for gases, and a Gas Chromatograph
Mass Spectrometer to test for organic residues in the soil.

The results for the Pyrotic Release and Gas Exchange Experiments suggested that non-biological processes only were operating. Meanwhile the Labeled Release Experiment produced ambiguous results: labeled gas was detected in the main sample and not in the control, but it was produced unevenly across the samples. The most significant result arguably came from GCMS, which found fewer organic compounds than *Apollo* had identified in the lunar regolith.

Viking was thus both a success and a huge failure. It worked, but the results seemed like a negation of human hopes for Mars. The most optimistic reading was that the evidence was inconclusive. More realistically, Viking seemed to close the question of life on Mars. NASA did not return to the planet for 20 years.

Following the Water

The debate surrounding life on Mars was rekindled in the 1990s thanks to a number of factors. In the first place, there were, and still are, scientists who questioned the settled interpretation of the Viking evidence. More significantly, evidence from NASA's Pathfinder mission began to indicate that there was water present in the Martian soil. Thirdly, studies on Earth of life in extreme environments was providing new insight into extremophile forms of life flourishing in improbable environments. The evidence of water was what really mattered. Follow the water – that has always been NASA's shortcut to finding life.

Pathfinder and orbital probes have enabled scientists to distinguish between two aspects of Martian water flow. The first is historic – there is clear evidence that Mars did once host substantial water, enough to form networks of valleys and to influence some of the minerals (chlorides and sulphates) now found on the surface. There is even some evidence that a briny sea covered a significant part of the northern hemisphere, although other studies suggest that Mars was arid and experienced only transient water flows.

It is important also to consider the Martian atmosphere and climate. There is some water vapor in the atmosphere, mostly produced by the melting of the northern polar cap, but it is in general very dry and dusty. There are seasonal variations in atmospheric pressure linked to a carbon dioxide cycle, but it is not clear that there was ever enough of this gas to create the surface pressure required for liquid water. Mars was once much warmer and therefore wetter than now, but this was a long time ago – the Martian Oceanus Borealis, if it existed, is dated to the Late Hesperian period.

The geologic history of Mars is divided into epochs, much as is done for Earth and the Moon. The earliest is the Pre-Noachian, from the formation of the planet until 4.1 billion years ago, and is followed by the Noachian, when the oldest extant surfaces were formed between 4.1 and about 3.7 billion years ago The Hesperian Period (named after Hesperia Planum) is dated from 3.7 to approximately three billion years and the Amazonian Period (named after Amazonis Planitia) from three billion years ago to the present.

The presence of water on the contemporary surface of Mars is highly disputed. Photographs acquired in May 2011 by NASA's Mars Reconnaissance Orbiter seem to show flows appearing in spring and summer on a slope inside Mars' Newton crater. Some 7 confirmed and 32 candidate Recurring Slope Linea (RSL) sites had been found by time of writing, all in the southern hemisphere on equator-facing slopes.

Fig. 7.4 Orbital imagery from the Mars Reconnaissance Orbiter combined with 3-D modeling shows flows that appear in spring and summer on a slope inside Mars' Newton crater (Credit: NASA/JPL-Caltech/Univ. of Arizona)

Images like this change our understanding of Mars once again. Are these streaks caused by water? The evidence is not conclusive, for we might be looking at flows of dry, granular material.

The greatest surface explorers of Mars have been the NASA rovers Spirit and Opportunity. Landing on the planet in January 2004, these remarkable instruments have achieved a huge amount of science and have also recaptured the public imagination for the Red Planet. Originally intended to last 90 Martian days, they exceeded every expectation and when Spirit, which became stuck in May 2009, was finally declared lost in May 2011, it had the kind of 'funeral' unimaginable for other instruments. Opportunity is still at work. On March 27, 2012, it had covered 21.35 miles and was at the north end of Cape York on the rim of Endeavor Crater.

If there was water, was there once life? Might there be life still in the briny flows, if that is what they are, on those equator-facing slopes? Spirit and Opportunity have established that areas of surface water did exist, but these sites have since been buried beneath dust that, through the action of frost, has cemented into soil and rock. There may still be liquid water in the depths.

Is There Life on Mars?

When it seemed that life had been found, not on the planet but frozen within a Martian meteorite, the world heard about it very quickly.

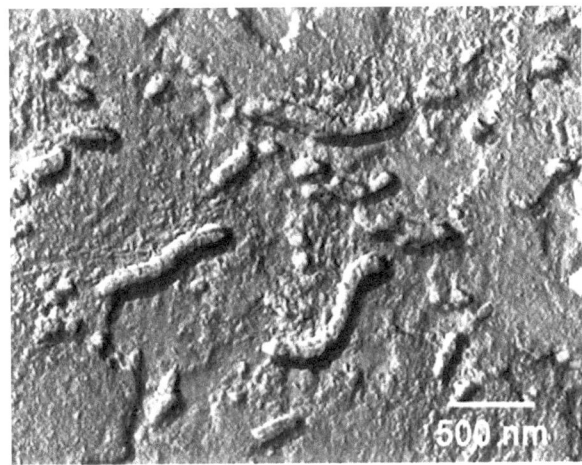

Fig. 7.5 Detail from within ALH84001 (Credit: NASA)

This is a slice of the most famous lump of Mars, the meteorite classified as Alan Hills (ALH) 84001, discovered at Alan Hills, Antarctica, in 1984. It became the object of intense study after it was claimed in 1996 that the tiny structures of magnetite, a mineral produced on Earth by water and soil bacteria, were microfossils, the first evidence of extraterrestrial life. This claim has been disputed, dropped and then revived, a study in 2009 concluding that "the biogenic hypothesis is still a viable explanation." The argument continues.

But ALH84001 tells us other things about Mars. It is very old compared to other Martian meteorites, and the mineral orthopyroxenite it contains has been reheated at least twice since initially forming from cooled lava some four billion years ago. At some time before it was blasted from the planet about 16 million years ago, it formed part of a wet environment. So if ALH84001 does not contain life, it does seem to confirm the watery hypothesis for a part of Mars' past.

Follow the water. The next NASA mission to Mars, the Mars Science Laboratory, seeks to do just that. On its way to the planet at the time of this writing in 2012, the mission includes a new rover, Curiosity, which is able to take extreme close-up images of the surface and vaporize thin layers of material with a laser before peering at the atoms this raises up. The goal is greater detail than has been possible before.

There are further tantalizing clues. Something on the planet is producing methane. There are various possible sources for this. It might indicate geologic activity or something organic, either of which would represent yet another revolution in our understanding of the planet. Recently, however, it has been suggested that the source is the degrading of meteorites subjected to high UV radiation on the planet's surface. Questions still remain to be answered about the planet, and the new surface rover is bound to answer some and then raise many more.

Fig. 7.6 Martian sunset (Credit: NASA/JPL/Texas A&M/Cornell)

The beauty of the Red Planet, though, remains a constant. So does the desire to grasp Mars' past history and present possibilities. The Sun, glimpsed in this image on May 19, 2005, by Spirit as it touched the rim of the Gusev crater, will surely never set upon our love affair.

Chapter 8

Future Suns

Observation: Orion Nebula M42, M43
Significance: What stars are made of and how they are formed
Science: Processes of cloud collapse and star formation, the Jeans Mass

Some time in the next few years, the Voyager spacecraft will leave the Solar System and enter into emptiness. Around 40,000 years later they will approach other stars. *Voyager 2* will pass within 1.7 light years of the red dwarf star Ross 248. *Voyager 1* will be near AC + 79388 at the same time. These first physical encounters between humanity and the stars will not be intimate. From the perspective of the spacecraft, both stars would be seen, were anyone still looking, to grow gradually in brightness and then fade.

From Earth, we glimpse moments in the lives of stars. As we have seen, the HR diagram helps us to see stars over the whole of their existence. In this observation, we are going to enter the stellar nursery. The Orion Nebula is one of the most studied and photographed objects in the night sky, and it helps us to explore the origins of stars. Seeing these stars answers some questions, and it poses many more. We will also take a look through the other end of the telescope, back at ourselves as it were. If we can see stars being formed, does that make any difference to us?

Observing the Orion Nebula

This observation begins with a return to Orion's belt, that beautiful line of the stars Mintaka, Alnilam and Alnitak. Hanging from the belt is Orion's sword, visible to the naked eye in most conditions. Follow the line of the sword down in binoculars,

M. Marett-Crosby, *Twenty-Five Astronomical Observations That Changed the World: And How To Make Them Yourself*, The Patrick Moore Practical Astronomy Series, DOI 10.1007/978-1-4614-6800-4_8, © Springer Science+Business Media New York 2013

and in its center you will find a star shining at a magnitude 4.2 but looking as if it is out of focus – 'woolly' was the word coined by some early observers. Now with the telescope, place the woolly patch in the center of the finderscope image. Then take a breath before you look at what you've found.

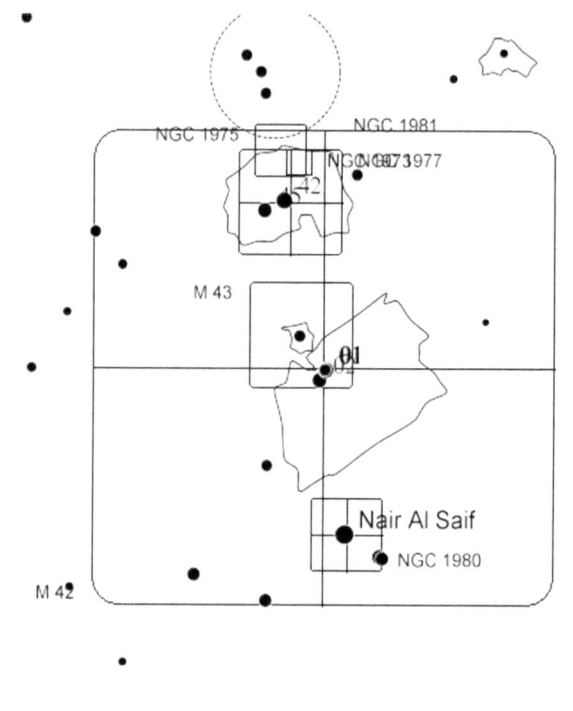

Fig. 8.1 The region around the Orion nebula (Cartes du Ciel)

A word on labels applied to deep sky objects. The Orion Nebula is commonly referred to by the number it received in the list of deep sky objects compiled by Charles Messier, meaning that this area of nebulosity was the 42nd item in his first list of 1,771. It is also NGC1976. Additional items identified by Dreyer between 1888 and 1894 have IC (Index Catalogue) numbers, and those spotted between 1895 and 1907, published as the Second Index Catalogue, have the prefix IC II. But many objects also have names attached to them: the North America, the Cat's Eye and so on. These are informal and range from the oddly accurate to the very imaginative.

Amateur astronomers can slide into a kind of rapture over the Orion Nebula. In any reasonable seeing, a 2- or 4-in. telescope will be able to separate four stellar jewels in a trapezium pattern. These are the stars Theta Orionis A, B, C and D.

A larger telescope with high magnification can reveal two more stars (Theta Orionis E and F), while very powerful amateur instruments may show G and H.

Of the four white Trapezium stars, Theta C is the brightest, at a magnitude of between 5.1 and 5.4 and belongs to spectral group O. It is both the youngest and the nearest of this type of star to us. It is thought to be not one star but three in close alliance, a main star some 31 times the mass of our Sun, a wide-orbiting companion 12 times its mass and, more uncertainly, a close companion star of our Sun's own mass.

Through our telescopes, Theta A and Theta D usually appear to be of the same brightness, but Theta A, a B-type star also called V1016 Orionis, is an eclipsing binary, a pair of stars the orbit of which crosses our line of sight from Earth. The result is that they eclipse each other. Theta A dims every 65 or so days. If it appears less bright than Theta D as you observe it – D is opposite A in the diamond pattern – then you are witnessing this eclipse. Theta B is also a binary system, formed from two mighty stars that eclipse each other every 6½ days for just less than 19 h. During this time, the magnitude drops from 7.9 to 8.7.

The Trapezium is at the heart of one of the richest areas of observation in the whole night sky, not only for its stars but for the nebulosity around it. When the Moon is in a quiescent phase, you might be able to sense the green tint caused by the Trapezium illuminating clouds of gas around and in front of them.

Set around the Trapezium is an area of nebula known as the Huygens Region, with an encroaching area of darkness to its left. This is the Sinus Magnus, Latin for Great Bay, which at the Trapezium end is cut by a tendril of light which, continuing the seashore metaphor, has the name Pons Schroerti, Schroeter's Bridge. These features are not straightforward to identify, but it is worth persevering. Schroeter's Bridge cuts the dark patch into a larger and a smaller area, as illustrated in this drawing made by the American astronomer George Phillips Bond (1825–1865) in 1863, here paired with an early photograph taken in 1883 by pioneer astrophotographer Andrew Ainslie Common (1841–1903). Common was, with Draper, the first astrophotographer to capture an image of an entire comet. Both he and Draper photographed the comet C/1881 K1 Tebbutt on June 24, 1881.

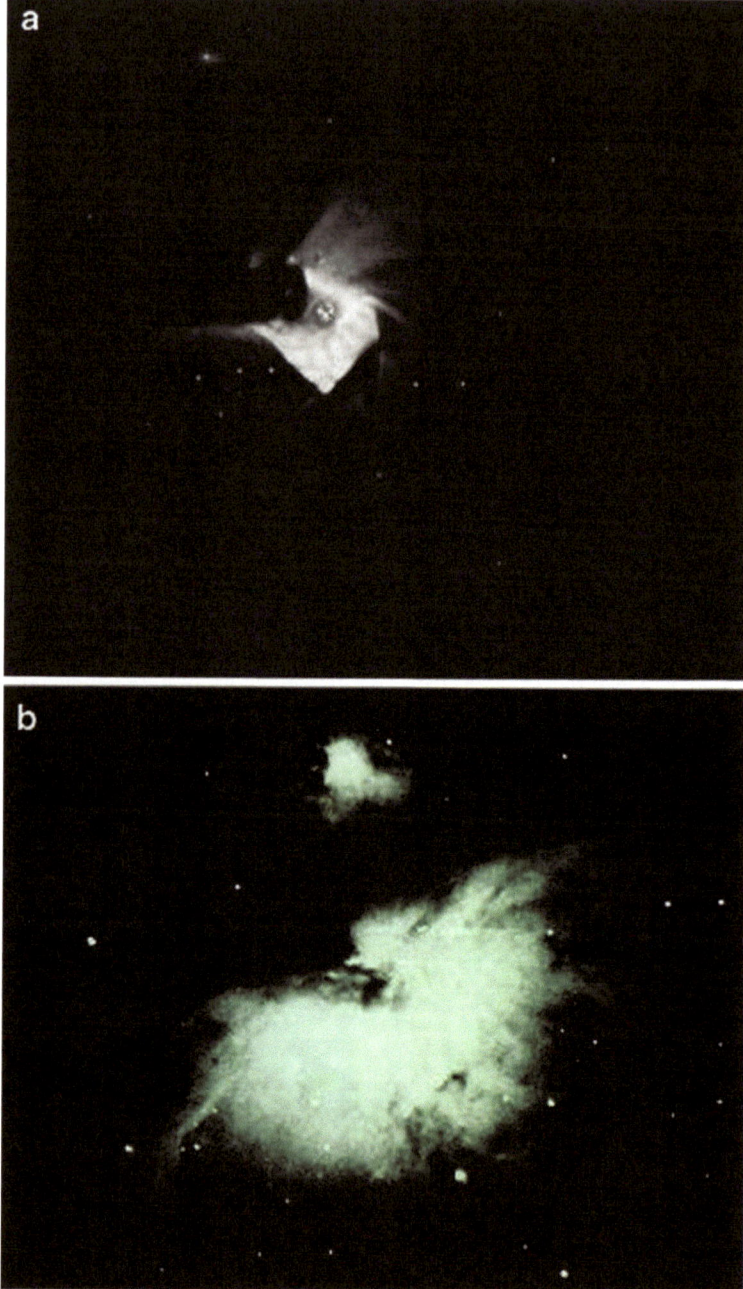

Fig. 8.2 An observational sketch of the Orion Nebula by George Bond made in 1863, compared with the same object photographed in 1883 by Andrew Common (http://www.springerimages. com/Images/Physics/1-10.1007_978-1-4419-5592-0_6-12)

From the Trapezium, use medium magnification to identify lanes of light cutting through darker areas. Above the main area, a river of deep dark separates two distinct bodies of nebulosity. This is a strip of dense cloud immune to starlight.

Messier distinguished between M42 and, directly above it, M43, which William Herschel described as "a circular glory of whitish nebulosity, faintly joined to the great nebula." Don't be tempted by too much telescopic power here. Go back to binoculars or a good wide-field eyepiece to gain a sense of the whole as well as parts.

Physically, M43 is not distinct from M42, its apparent separation another result of the dust lanes intervening between the nebula and our line of sight. The nebula is illumined by the star NU Orionis shining through it. (Note that the NU Orionis is pronounced '*N-U Orionis*.' The letters NU are a two-letter variable star catalog reference and not the Greek letter nu (v). The star is at least a binary system, of which the principal is a B-type star some 19 times the mass of the Sun.) The star's influence over its 'local' dust seems pronounced because the dark band we have observed stops the light of the Trapezium stars from leaking. It may be that other scattered light from within the Huygens region may also promote M43's particular shine. The 'comma' shape of this nebula is reasonably clear, and through a large instrument it is possible to determine some of its coastline-like features, especially its bays.

There is a huge amount to observe and explore in the Orion Nebula and its surrounds. A good telescope will reveal more patterns of nebulosity above M43, a series of bright patches interlaced with darkness sometimes called the Running Man Nebula, NGC1977.

The greatest prize, however, for amateur observers in this part of the sky is the Horsehead Nebula, a dark patch carved into a long streak of nebulosity not far from Alnitak. In the best observing conditions and with a large aperture telescope (opinions vary as to how big it has to be – and some wild claims are made – but 8 in. is probably the minimum) edge northeast of Alnitak to NGC2024, the Flame Nebula. Worth observing in itself, the Flame is a useful test – if it is clearly visible, then the atmospherics are on your side. Move to the south of Alnitak in as near a straight line as you can from the bright star embedded in NGC2024, looking for a 7.8 magnitude star with its own patch of nebulosity. This is NGC2023.

You are now close to the Horsehead. The only star of comparable brightness in a field of view with NGC2023 on the left side is HD37805, an outrider to the strip of Horsehead nebulosity. It is difficult to make this out, and achieving the right contrast can require either a filter or very high magnification, with all the challenges that can bring. Ease away from the star and get settled, for the nebula's glow is faint and easily swallowed by other sources of light. It takes time to see that it's there. IC434, the nebula, is extended and thin. The dark intrusion of the Horsehead, Barnard 33, is a cloud of dust about half way along its length as you move from the Flame Nebula.

There's no doubt that this is hard. The result will not be that you obtain a Hubble-type image in your eyepiece. But amateurs can achieve remarkable pictures of the Horsehead, as this example shows.

Fig. 8.3 The Horsehead Nebula (Barnard33 in bright nebula IC434), taken with an Orion ED80 and Atik 16HR. Fourteen Exposures of 120 s were stacked using Astroart 3.0 (Credit: Peter Campbell-Burns, http://www.farnham-as.co.uk)

A Chaos of Future Suns

Now let us try to understand what we've been seeing.

We can start with the words of the American astronomer E. E. Barnard (1857–1923), who prefaced his 1919 catalog with these words: "It would be unwise to assume that all the dark places shown on photographs of the sky are due to intervening opaque masses between us and the stars. But in a considerable number of cases, no other explanation seems possible."

In other words, darkness is not nothingness. The areas around the Trapezium stars in M42 and NU Orionis in M43 are full of dust particles in the interstellar medium, a diffuse swirl of tiny grains mixed with gas. The actual amount of material is very meager. Most of this gas is hydrogen, the simplest and most abundant element in the universe.

This interstellar material forms clumps. If you think back to the observations we have made, it's clear that there must be differences of density within the interstellar medium. We noticed how the light of a star such as NU Orionis was bounded, unable to penetrate the mass. Gatherings like this are key to unlocking one of the most exciting, and uncertain, of modern astronomical discoveries, the mechanism by which stars are formed.

Before we do so, we need to understand more precisely what astronomers mean by nebula. The word *nebula* is derived from the Latin for a cloud. Its early English uses were distinctly unastronomical – a nebula was a cloudy patch in a urine sample. For a time, familiar *nebula* was used alongside *nubecula* to mean the rim around a sunspot as well as any cloudlike object in the sky. *Nubecula* survives as the name for the Magellanic Clouds of the southern hemisphere, while *nebula* has emerged as the general name for the feature.

Distinctions are made between different types of nebulae, beginning with those that, like Orion, glow with their own light. Only the most vigorous stars produce sufficient photons to achieve this, so emission nebulas, as they are called, mark some of the most dynamic areas in the sky. A celebrated example is the Eagle Nebula, M16, in the constellation Serpens.

A subclass of emission nebulas are those formed after stars explode. The gas remaining after a supernova spreads into delicate filaments or threads. We will meet one of these when we encounter the Crab Nebula in a later chapter.

Also in this group of emission nebulae are the confusingly named planetary nebulae, such as the Dumbbell (M27) in Vulpecula, the faint fox constellation beneath Cygnus. As a Sun-like star enters its red giant phase towards the end of its life, it can expel its atmosphere, creating a ring of gas thrown into space. Planetary nebulae have nothing to do with planets save for a serendipity of observation. William Herschel used the label because their appearance reminded him of the planet Uranus.

Another broad class of nebulae glow by reflecting scattered starlight, dispersing it such that we see a characteristic blue glow. We have already met a reflection nebula when spotting the Pleiades or Seven Sisters, M45.

There are also dark nebulae, clouds so dense that no light passes through them. We can only see these when silhouetted against brighter backgrounds. The Horsehead is just such an object.

The Orion nebula is an emission nebula, the type most associated with star formation. It is a dense as opposed to a diffuse cloud. The hydrogen within has been ionized to create what is called an HII region. It is these dense clouds that are the birthplaces of stars. Observationally, HII regions occur around young stars – those we have been observing are about 10,000 years old, children on the interstellar scale – so the connection seems sound. But how does it happen?

We can start with mathematics and the work of Sir James Jeans (1877–1946), whose modeling of clouds led to the prediction that a dense, cool cloud would tend to contract upon itself under the influence of gravity, exerting a stronger inwards force that the outward force of pressure derived from motion. According to Jeans, this contraction would occur if the mass of the cloud was sufficiently large, the so-called Jeans mass or Jeans criterion. He had identified the triumph of gravity over pressure.

(Sir James Jeans was a creative astronomical theorist, and never more so than in the model he proposed for the creation of planets. He suggested that our Solar

System resulted from a massive star passing our Sun and exerting such a tidal pull upon its mass that a filament of solar material broke away. This bit of Sun then separated into blobs of proto-planet).

The problem is that not all clouds, even dense ones, actually collapse. They require a trigger. The mechanisms at work here are not clear. Shockwaves from supernova explosions or the close approach of stars and galaxies might set the contraction process in motion, but we are not certain.

What happens to our collapsing cloud? As contraction continues, so the energy of gravity is converted into kinetic and finally thermal energy. Tiny particles move faster, and they collide. Temperatures rise such that, after just a few thousand years, this swirling, hot mass can be called a protostar.

This protostar is still some way off the familiar areas of the Hertzsprung-Russell diagram, but so-called Hayashi tracks have been plotted by which contracting clouds of different masses might become the different sorts of stars. We cannot see this happening, but we can glimpse some of the moments during the transformation from protostar to star.

Fig. 8.4 The Hubble telescope's infrared image of the Orion Nebula, with the visible-light view on the *left* (Credit for the NICMOS image: Rodger Thompson, Marcia Rieke, Glenn Schneider, Susan Stolovy (University of Arizona); Edwin Erickson (SETI Institute/Ames Research Center); David Axon (STScI), and NASA. Credit for theWFPC2 image: C. Robert O'Dell, Shui Kwan Wong (Rice University) and NASA)

Here is the nebula as you cannot see it through your telescope. The right-hand infrared image witnesses to the fury of star birth with the massive Becklin-Neugebauer object at its center. The blue streaks around it, or hydrogen, being pumped out by young stars are locked in the dust.

The end of the infancy of stars comes when the temperature in the core reaches the point at which nuclear fusion becomes possible. We have seen something of how this works when looking at our own Sun. And long after we have finished watching, the Orion Nebula will change. The Trapezium stars will disperse the gas clouds that made them, leaving behind a bright cluster much like the Pleiades. The Horsehead Nebula will probably disappear, so there is only about 100,000 years or so to get to see it. Over their aeon-long lives, these stars will pass along the HR line until they, in their turn, run out of nuclear fuel and fail.

There are uncertainties with many aspects of this theory of stellar birth. Although star creation is the fundamental astronomical process, the current theory contains some awkward steps. The formation of cosmic clouds, dense or otherwise, is not well understood, and the trigger mechanisms for their collapse are uncertain. Our theory is poor at explaining the birth of massive stars, which would become so luminous so quickly as to disrupt their own formation. New observations may help to confirm or refute these and other aspects of the theory.

The idea that stars have a life from birth to death is very new. Classical philosophies viewed the stars as eternal, their perpetual shining bounded only by divine creation and the end of all things.

Which was a problem when new stars were seen to appear. These events, which we now call novae or supernovae, result in previously invisible stars flaring into a brief brightness as they die. This is the modern explanation. We will observe supernovae and their remnants in another chapter. But they have been visible throughout observing history, and some observers were brave enough to accept that they were new stars and propose theories as to how they were formed.

One such was Francesco Maurolyco (1495–1575), an Italian Benedictine monk who, in 1577, observed a new star in the skies above Messina. He wrote about it, speaking of "an unusual sign more marvelous than a comet," a star where no star had previously been seen. It was wonderfully bright, shining more clearly than any of the first magnitude stars.

But what could cause this? There was no room for new creations of stars in the Aristotelian synthesis of his time. But Maurolyco was prepared to think afresh. He proposed a method by which a star made be: "Made in the heaven out of the aggregation of the brilliance of planets and the rest of the stars."

A star was born, then, out of an excess of brightness. It is a beautiful idea. And although Maurolyco's idea does not find its way into modern accounts of stellar formation, it is something to take with you on nights viewing the Orion Nebula, as we look into the fire of new stars.

The discovery of the Orion Nebula is credited to various seventeenth-century astronomers: Nicolas Claude Fabri de Peiresc (1580–1637) observed it in November 1611, as did the Swiss Jesuit Johann Baptist Cysat (1587–1657) in 1618. Evidence from the Mayan civilization in Mexico suggests that the nebula may have been seen and known from there much earlier, for tradition ascribes flames to that part of the sky described as Orion's sword. Its Mayan name refers to smoke rising from burning incense, set between three stones in a hearth. It's hard in the absence of other records to go further than the suggestion that the Mayans had observed the nebula, but given what science now tells us about such regions, a place of fire is not so far from being accurate.

Galileo did not spend much time studying Orion through his telescope before the publication of his *Starry Messenger*, but he did note many new stars in the constellation, stating that: "To the three in Orion's belt and the six in his sword that were observed long ago, I have added 80 others seen recently."

His drawing of the sword and belt does not show the nebula, but he did observe and sketch it in later years. We can date precisely when and where he turned his telescope towards it – the night of February 4, 1617, in the sky above Bellosguardo – because his observation is recorded in his posthumous *Analecta Astronomica*. He drew an outline of the Trapezium as well. But this never made it into print, and so the revelation of the Great Orion Nebula's wonder was left to others.

The Orion Nebula occupies a special place in the history of astrophotography. The first photograph was taken by Henry Draper (1837–1882), son of a distinguished New York doctor who himself had been an early photographer of the Moon. Raised in a household of scientific endeavor, Henry Draper, himself a doctor, used dry plates and an 11-in. Clark refractor to record images of the nebula in September 1880 and again in March 1882 just before his death. This first image is still evocative and powerful.

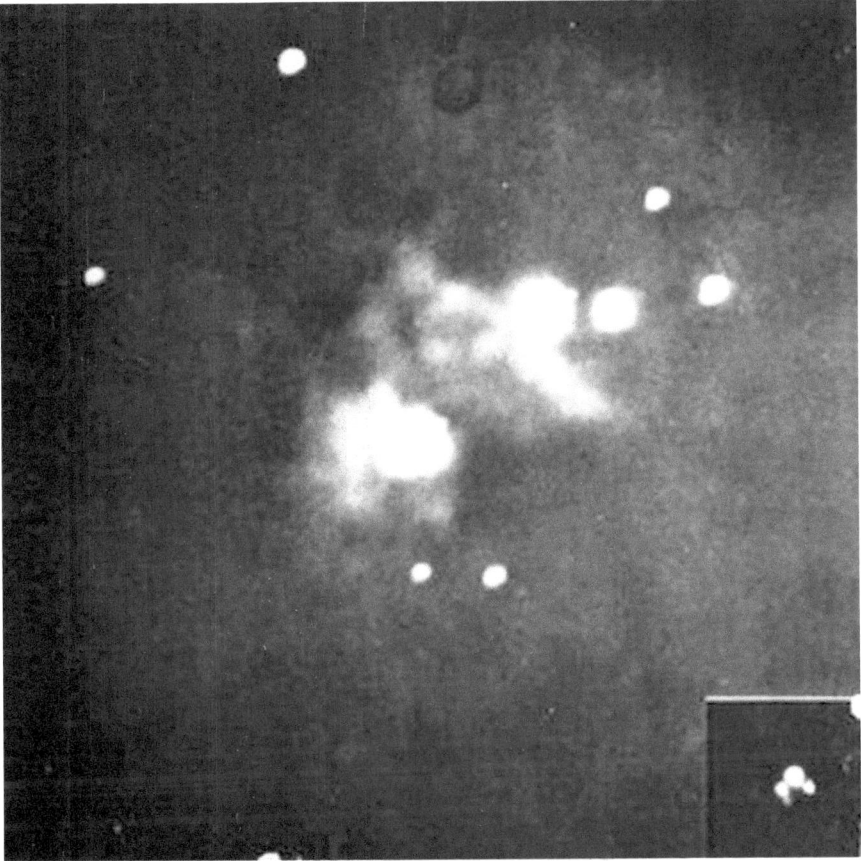

Fig. 8.5 The first deep-sky image, a photograph of the Orion Nebula taken by Henry Draper in 1880 (http://www.springerimages.com/Images/Physics/1-10.1007_978-1-4419-5592-0_6-4)

It was photographs in Draper's tradition, albeit with better equipment, that enabled E. E. Barnard to create his enormous archive of nebula photography during the first part of the twentieth century. This provided the raw material by which Edwin Hubble was able to distinguish between the types of nebulae.

The nebula was among the first deep sky objects photographed by astronauts in space. Pete Conrad and Richard Gordon captured it from the *Gemini 11* probe in September 1966. The Hubble Space Telescope is the latest, and thus far the finest, source of such images.

Launched in April 1990, Hubble has been able to escape the atmospheric limitations imposed on ground-based imaging, and once a primary mirror fault was corrected by the Space Shuttle service mission of December 1993, it produced images that effected fundamental changes in scientific theory. The science section of this chapter alone would be absolutely different without Hubble's detection of protoplanetary nebulae around young stars.

Fig. 8.6 Astronaut F. Story Musgrave, anchored on the Space Shuttle Endeavor's robotic arm, prepares to be elevated to the top of the Hubble Space Telescope during Hubble's first servicing mission in 1993. Astronaut Jeffrey Hoffman, inside the shuttle payload bay, assists Musgrave (Credit: NASA / STScI)

How is a Hubble picture taken? Apart from the complexities of bidding for the project and the time, scientists working with technicians have to remember certain rules. One is now familiar – Hubble, just like us, would be irreparably damaged if it was pointed towards the Sun. Indeed, its protective shield drops automatically at an angle of 20° to the solar glare. The telescope uses guide stars to fix its position and then takes the image using one of its several different cameras.

Hubble is a telescope in motion. It is in orbit around Earth at a speed of 5 miles a second, collecting light on a 94.5 in. mirror. Four antennae then send the data back to the Goddard Space Center. The telescope transmits something like 120 gigabytes of science data in a week. This compares, for those who still like their books, to more than 3,000 ft of new shelving.

Almost the only objects out of Hubble's reach are the planet Mercury, too close to the Sun, and the surface of Earth. Hubble is moving far too fast to picture anything of us save for a blur.

The list of its firsts is a catalog of advances in astronomy, from exoplanet atmospheres to its deep field images of impossibly distant galaxies. Hubble has identified two moons of Pluto and watched the disintegration of a comet in the atmosphere of Jupiter. There will be no more servicing of Hubble now that the space shuttle fleet has been retired, but the photography of astronomical objects from space will continue with the James Webb Space Telescope (JWST), due to be launched in 2018. At time of writing, NASA has also been gifted two space telescopes at least as powerful as Hubble by the U. S. National Reconnaissance Office, but there are no funds available to develop a space mission to launch them before 2024.

Beware of Pictures?

It is hard to overstate the importance of pictures of the stars. Photographs allow for the recording of greater detail, permitting long exposures and light collection on the plate or film or digital receiver. The power of such images is evident in the Hubble pictures we have seen thus far. And they have had cultural significance as well – for good and also less good, pictures of the heavens have changed our relationship with them.

One consequence of the ubiquity of images is that, compared with the wonder of the Hubble images, the sight of the Orion Nebula through your telescope might feel disappointing. Hubble pictures are colorful, sharp and immediate. They are also easy to paste onto laptop screens, in any weather and at any time of the day. They shrink onto postcards and they become malleable. We can Photoshop the Cat's Eye Nebula onto the image of our kitten.

So why not put away the telescope and become Internet astronomers? Because direct experience matters. The computer screen and image processing, however wonderful, act as a veil between the seer and nothing seen. It is a point incisively expressed by the thinker Walter Benjamin (1892–1940), who wrote: "To perceive the aura of an object we look at means to invest it with the ability to look at us in return." Recall the awe-inspiring fact that the 1,500 year journey of light from the Orion Nebula has ended up on your own eye.

To look at something on a website costs so little in terms of time that we all graze, flicking between wonders almost carelessly. Whereas to go outside, prepare the telescope, hunt for an object, lose it, get cross, go back inside for coffee and then find it again, represents an investment. That means it brings a return.

This is not an argument for saying that all technology is bad. It is to say that we should still look, and for ourselves, at what we can. Susan Sontag writes that photography will tend to bring "knowledge at bargain prices" and "a semblance of wisdom." It is much as Gabriel Josipovici argues in his study of perceptions, *Touch* (Yale University Press, 2011, p. 17): "The very abundance and proximity of masterpieces...tend to deprive each of their aura."

Aura. That is a good place with which to end this observation. The milky vision of Orion, perched in the constellation's sword, that trembling Trapezium of young stars, are places none of us will ever go. Nevertheless, we can see it for ourselves. That is surely an awe-inspiring thought.

Chapter 9

Bright Dog and Dark Companion

Observation: Sirius and the constellation Canis Major
Significance: Distances to the stars
Science: White dwarf stars, parallax

Our ninth observation continues our journey through the stars. To begin with, it is straightforward – we are going to find the brightest star in the night sky. But don't let the ease of finding it lull you into thinking that Sirius is nothing much. It is the gateway to a fascinating night under the sky and takes us from the brightest star to invisible companions, by way of an accurate measurement of the extent of the sky.

How to Find Sirius

After the Sun, the Moon and Venus, Sirius is consistently the brightest object in the night sky, though sometimes outshone by Jupiter and Mars. It is often the only star to make its way through the sodium glare of a city street.

To identify Sirius, find the three stars that form Orion's belt and follow their line southwards to a dazzling, bright blue star that stands a little ways off from any other. This is Sirius, the Dog Star, shining at a magnitude of −1.46. Sirius stands alone and looks splendid to the naked eye, while through binoculars, it offers an extraordinary palette of colors, flashing blue and green and also red. Try to spot the red, for we will return to red Sirius later.

Sirius is the alpha star of the constellation Canis Major, the greater of Orion's pair of hunting dogs. Its brightness is the consequence firstly of proximity. At 8.61 light years from us, Sirius is one of our nearest stellar neighbors. Its intrinsic luminosity

M. Marett-Crosby, *Twenty-Five Astronomical Observations That Changed the World: And How To Make Them Yourself*, The Patrick Moore Practical Astronomy Series, DOI 10.1007/978-1-4614-6800-4_9, © Springer Science+Business Media New York 2013

is also considerable, shining 25.4 times brighter than the Sun. It is a classic example of an A-type star, a well-represented group among naked-eye stars.

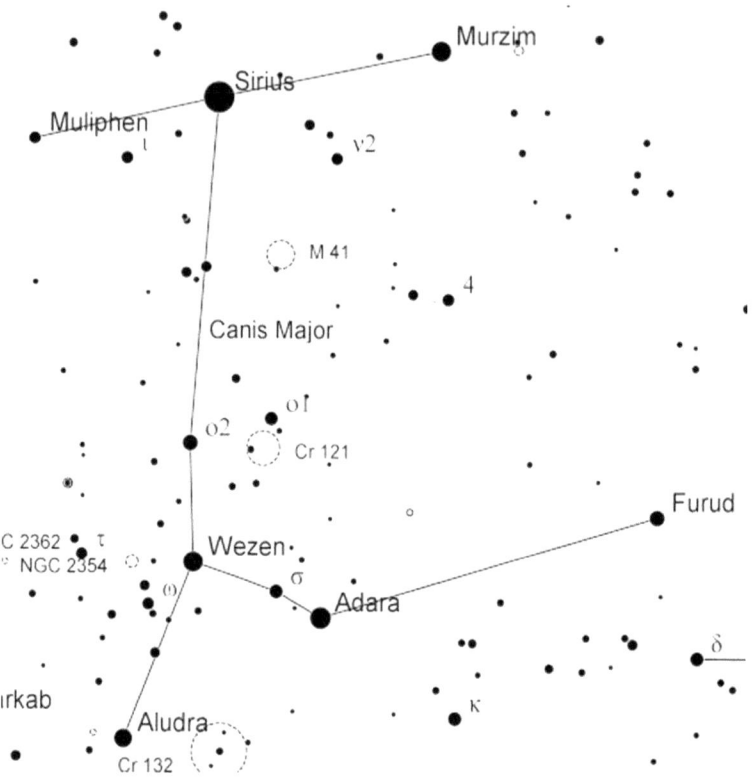

Fig. 9.1 The constellation Canis Major, with Sirius at its head (Cartes du Ciel)

The constellation of Orion's greater hunting dog needs some imagination to envision. But Canis Major contains many astronomical treasures observable with binoculars or a telescope.

Immediately to the west of Sirius lies Beta Canis Major, the hot B-type star Mirzam. Although seemingly less bright than its mighty companion, Mirzam is in fact the larger, a bright giant 60 times more distant from Earth than Sirius and shining much more brightly than the Sun. Recent measurements suggest that, had we been looking at the sky some 4½ million years ago, Mirzam would have appeared significantly brighter than Sirius, more brilliant even than Venus. It is a complex variable star of a type called Beta Cephei, which we will meet again in a future chapter.

We do not know what the name *Mirzam* means, but it was applied in Arabic charts to several stars whose rising preceded very bright stars. There are Mirzams near Betelgeuse and Procyon. The name may well refer to this phenomenon.

Sirius and Mirzam form a short, straight line. If you imagine a right-angled triangle with this line as one side and turn 90° away from Orion at Sirius, you reach a true supergiant, the F-type star Wezen, Delta Canis Majoris, which has a radius some 215 times greater than our Sun. Its companion to the south is Epsilon Canis Majoris, or Adhara, a beautiful blue star in the B spectral group. Adhara carries the distinction of being the brightest known source of emissions in the extreme ultraviolet wavelength. In November 1996, it was studied through a spectrometer carried on the Orfeus-Spas II platform, enabling a close study of the star's photosphere.

No star, however, compares to VY Canis Major, an oxygen-rich M-type red supergiant which, with a radius 1,420 times that of the Sun, is the largest star we know of in the sky, exceeding both the Pistol Star and Eta Carinae. If it were in our Solar System, its surface would extend as far as Mars or even Saturn, depending upon where the 'edge' of the circumstellar envelope really lies. VY Canis Major is an extraordinary chemical laboratory containing a rare mixture of chemical elements. Its extended atmosphere has been recently studied through the VLT array and an envelope of water and carbon monoxide found close to the photosphere.

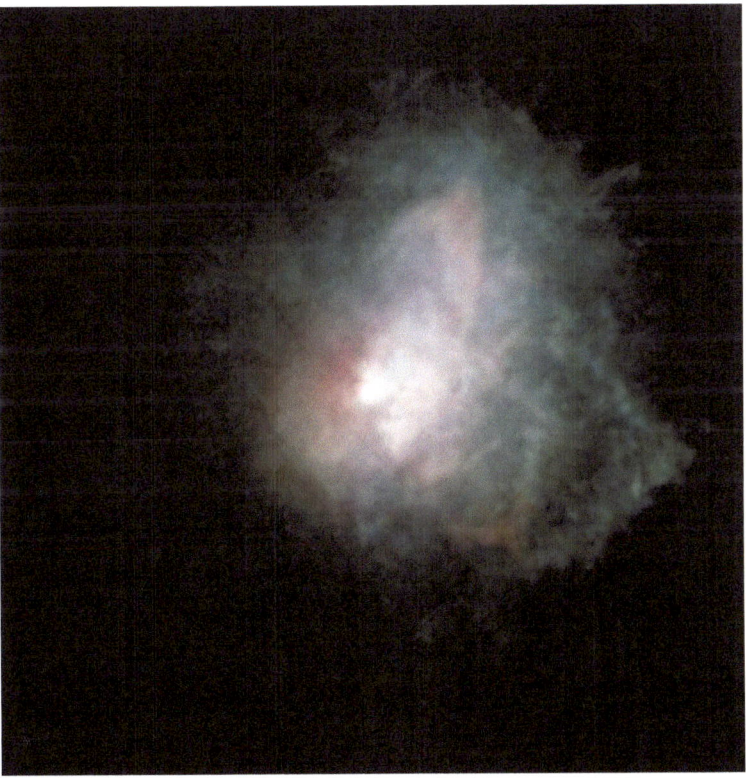

Fig. 9.2 VY Canis Majoris in polarized light (Credit: NASA, ESA, and R. Humphreys (University of Minnesota))

Its size does not make VY Canis Major easy to see. The star lies above Wezen, making it a kind of 'third leg' of the dog. But it is more than 5,000 light years from Earth and little more than the tiniest point of light.

Returning to the line we made from Sirius to Wezen and retracing our steps towards the alpha star, we pass close to the open cluster M41, which lies almost directly beneath Sirius and may be the longest known of all the open clusters in the sky, quite possibly mentioned by Aristotle in his *Meteorologica* of 325 BCE. "One of the stars in the thigh of the dog has a tail, though a dim one." This line is thought to refer to M41. The philosopher also seems to have understood how averted vision works. He says "If you looked hard at it, the light became dim, but to a less intent glance it seemed brighter."

M41 is one of few open clusters that can be seen well with binoculars. If you can keep the image steady, it is possible to distinguish as many as a dozen of its constituent stars, whereas with a small telescope, 30 or more stars appear out of a cluster gathered into a 20 light year diameter group. M41 is dominated by red giants of which the central star HD 49091 stands out clearly. This is a really rewarding target for observation and for astrophotography, as revealed in this image.

Fig. 9.3 Open Cluster M41, GSO RC 10″ F8 2,000 mm – astrograph Ritchie-Chrétien and ATIK 4000LE with Baader LRGB 2″ filter, taken 28 March 2012 at Pragelato, Turin, Italy (Credit: Leonardo Orazi, http://www.starkeeper.it/)

The final stop on our tour of Canis Major is the binocular double star Furud (Zeta Canis Majoris). Furud forms a triangle with Sirius and Wezen, and it lies almost directly beneath Mirzam. With binoculars it is possible to distinguish Furud's primary star from its secondary a little to its north. It is one of those 'there and not there' pairs, requiring averted vision, as Aristotle recommends.

What We Are Seeing

As we have seen, Sirius itself is a hot A-type star only 8.6 light years from Earth. It is not alone. Nestling within its brilliance is a companion star, Sirius B. It is the nearest white dwarf to us in the sky.

It is worth reminding ourselves of the distinction between two types of paired stars – those that appear to be close together and those that actually are. The first group are called optical doubles, stars that seem intimate from our standpoint as observers but have no relationship in fact. Optical doubles are an effect of line of sight. They are quite different from true binary stars, which interact with one another while orbiting a common center. This was a division established by Sir William Herschel when he published in 1802 his catalog of 500 nebulae and clusters:

> If a certain star should be situated at any, perhaps immense, distance behind another, and but very little deviating from the line in which we see the first, we should then have the appearance of a double star....If two stars should really be situated very near each other... then they will compose a separate system, and remain united by the bond of their own mutual gravitation...a real double star.

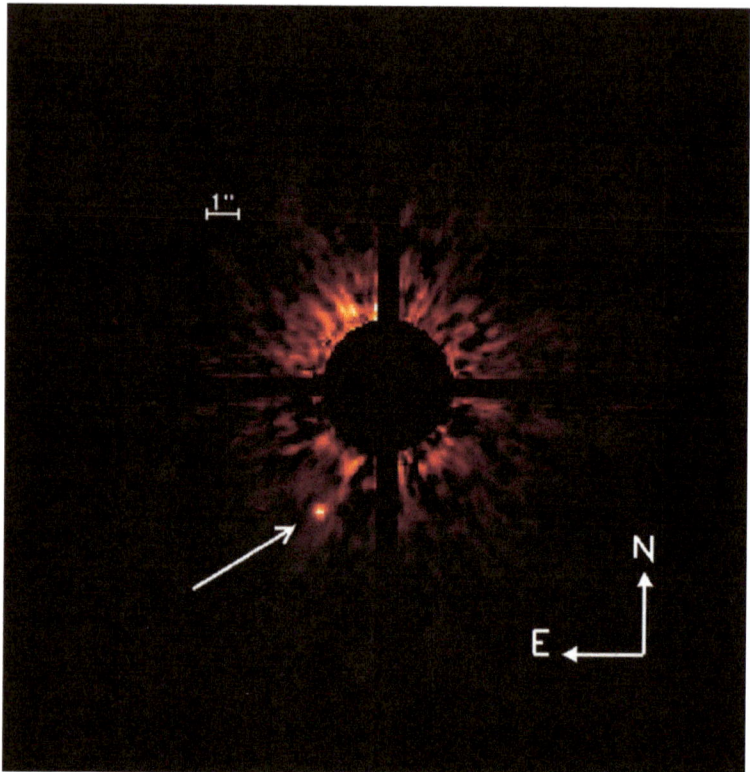

Fig. 9.4 ADONIS high-contrast image, with Sirius B marked with an *arrow* (Credit: Bonnet-Bidaud and Pantin, Sap-CEA/ESO)

Here is an image of Sirius as you will not see it. The brilliance of the main star is spectacular. Below and to its left is the dot of Sirius B, the white dwarf star.

White dwarf stars – their name tells us two things. First, they are small. White dwarfs are often no bigger than the diameter of Earth. Sirius B measures approximately 12,000 km across, just a little smaller than our planet's 12,756.32 km. As a consequence, their luminosity is low; there are no white dwarfs that can be seen with the naked eye. (Keep in mind that luminosity depends upon two factors, temperature and size. The relationship is expressed in an equation $L = 4\pi r^2 \sigma T^4$, where r is the radius of the star, T its temperature and σ the Stefan Boltzmann constant.)

The second thing we know is that white dwarfs are hot, with surface temperatures in the range of 5,000–25,000 K. They belong at the end of the stellar evolution sequence and gather towards the bottom of the Hertzsprung-Russell diagram. White dwarfs have run out of their nuclear fuel, and their cores are therefore immensely dense. Sirius B has a gravitational field some 350,000 times that of the surface of Earth.

The density of a white dwarf star keeps it in existence. Under these immense pressures, an interior structure called a degenerate electron state comes into being. This is stable so long as the white dwarf stays within a limit of 1.4 times the mass of the Sun, a limitation upon possible mass predicted mathematically in 1930 by the Nobel prizewinner Subramanyan Chandrasekhar (1910–1995).

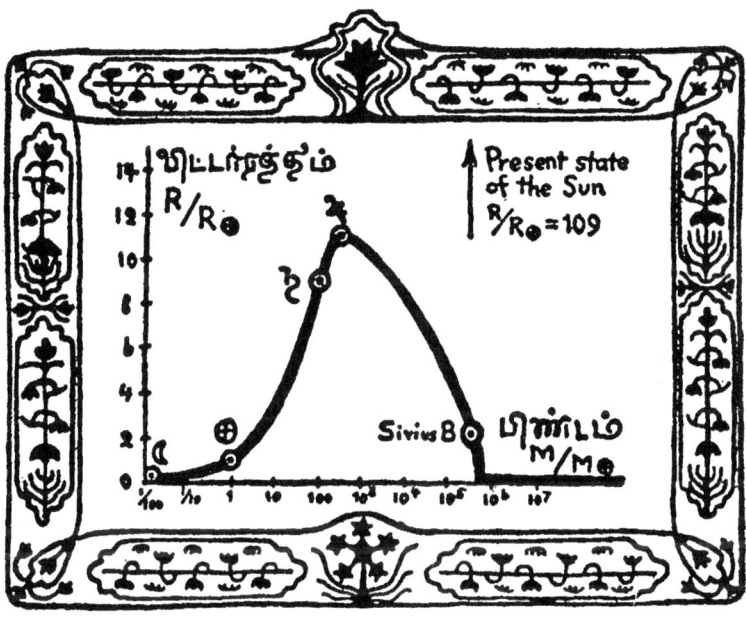

Fig. 9.5 The Chandrasekhar Mass-Radius Graph

Chandrasekhar's mass-radius relation is expressed here in the original graph. The labels are in Tamil, his first language. You will see where the Moon, Earth, Saturn and Jupiter all fall on the line, a long way before it reaches Sirius B.

And beyond? The line goes flat. After the Chandrasekhar limit, we enter upon the territory of neutron stars and black holes, states of catastrophic collapse that we +will return to, and even observe (sort of), in a later chapter.

Chandrasekhar described Sirius B as "highly underluminous," reminding us that the astronomy of the less visible is just as important as the astronomy of the bright. Sirius and its companion are a study in light and gloom. Both have mattered to the history of science.

Sirius posed questions to the astronomers of history. If Sirius is the brightest star, is it also the nearest? If all stars shared the same intrinsic luminosity, the answer would have to be yes. So just how far away is Sirius?

This was the question that the astronomer Christian Huygens (1629–1695) set out to answer, using no more than a thin disc of brass punctured with holes. He let the Sun shine through the holes, magnifying its image until the brightness he saw on a screen was equal to that he remembered from observing Sirius. The diameter of the image was 27,664 times that of the hole, so Huygens concluded that Sirius was 27,664 times the distance of the Sun from Earth.

It was not the most accurate of methods, depending as it is on Huygens' memory of how bright Sirius might be. But he reached a dizzying number all the same. The Scots mathematician James Gregory (1638–1675) also compared Sirius's brightness with that of the Sun by making use of Saturn, and he came to an even greater figure. Finally Sir Isaac Newton (1643–1727) worked out in 1728 that Sirius was as much as one million astronomical units from Earth.

The point was very clear – the scale of the stars was immense to the point of being inconceivable, and Earth was by that vastness made smaller. Sirius was the teacher in that humbling class.

However, the numbers achieved were very varied, and the methods of measuring how far away the star lay were all of them in different ways unsatisfactory. The secure means of obtaining these distances was parallax, the careful measurement of the star's position with respect to more distant stars behind it and a second measurement 6 months later when Earth is on the opposite side of its orbit. The shift is very small, but it is enough to allow for the measurement of the parallax angle and so the distance.

Parallax was a challenge. "It is a difficult matter," Tycho Brahe had written, "and one that requires a subtle mind, to try to determine the distances of the stars from us, because they are so incredibly far removed from the Earth." Galileo had known this and had sought parallax from Mizar and Alcor. But it required far more powerful telescopes than he had to achieve the goal of learning how far away the stars really were.

Several nineteenth-century astronomers sought out parallax. Wilhelm Struve (1793–1864) persuaded the Prince of Dorpat to let him purchase Fraunhofer's 9½ in. Great Reflector and focused this near perfect instrument on Vega, making 17 observations and a parallax measurement of one eighth of an arcsecond. In July 1837 he reported this to his friend Freidrich Bessel (1784–1846), who was using the Konigsberg heliometer atop its 44-foot tower to study 61 Cygni, a true double with motions that could be compared against each other to determine the tiny angle created by Earth's rotation.

In December 1838, Bessel told the astronomical world that he had achieved the measurement. The angle was 0.314 of an arcsecond, placing 61 Cygni some 660,000 AU, or 10.4 light years, away. Two months later, the first Astronomer Royal for Scotland, Thomas Henderson (1798–1844), announced that he had measured the parallax of Alpha Centauri.

The 61 Cygni pair are K-type stars orbiting each other with a period of about 722 years. They lie not far from Deneb (Alpha Cygni) on a line between it and Zeta Cygni on the outstretched arm of the Cross asterism towards Pegasus.

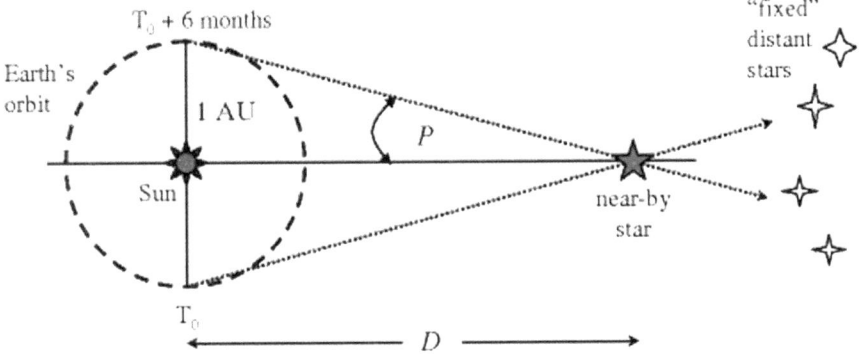

Fig. 9.6 Stellar parallax due to Earth's motion around the Sun. Angle P is the angle of parallax, and the distance to the star is given in terms of Earth's orbital radius, defined as 1 astronomical unit (AU) (http://www.springerimages.com/Images/Physics/1-10.1007_978-1-4419-5668-2_6-1)

Here was the scale of the universe and the final proof of the Copernican model. Galileo and many others had sought for this great prize. An astronomer we will meet in a later chapter, Ole Römer, spent many years seeking out parallax and wrote that: "I was afraid that, by pursuing such a tiny little thing, which almost escapes the sense, I would expose my work to scorn or mistrust."

Bessel had not finished with tiny little things. He next turned towards Sirius. He came to the conclusion that it was changing in such a way that could only be caused by the presence of an unseen companion. He wrote to Humboldt: "I adhere to the conviction that Procyon and Sirius are genuine binary systems, each consisting of a visible and an invisible star. We have no reason to suppose that luminosity is a necessary property of cosmic bodies. The visibility of countless stars is no argument against the invisibility of others."

We see here an important step being taken in the understanding of the stars. The telescopic revolution extended the realm of the visible. Bessel perceived that astronomy would also be concerned with the invisible, and that human sight could not encompass the entire cosmos.

If Sirius had an invisible friend, what was it? The answer came on the night of January 31, 1862. when Alvan Graham Clark of Harvard was working with his father to test the color correction of a new 18.5-in. refractor telescope. He is said to have called out, "Father, Sirius has a companion!" This discovery was confirmed and then published by George Phillips Bond, who repeated Clark's observation using the Harvard 15-in. telescope.

Others were watching Sirius as well. While the Clarks were testing their new lenses, the Paris Observatory was beginning to use its 32-in. reflector, almost immediately focusing it on Sirius. The news of a confirmed companion brought recognition and success to Alvan Graham Clark and confirmed the power that modern telescopes might give to the astronomer.

This was not the end of the mystery of Sirius. Around 1930, a third star was reported in the system, provisionally named Sirius C. Observations using corona-graphs to block the glare of Sirius A searched in 2008 for anything up to 30 times the mass of Jupiter, repeated in 2011 at an even smaller scale, but both failed to find any point source other than Sirius B. This weakens considerably the likelihood of there being a Sirius C, but it is still not impossible that another star is hiding in the brilliance of Sirius itself.

The Dog-Day Star

Repeat your observation of Sirius and note the colors that you see. The star seems to flicker and change.

One color that does not endure is red. Yet ancient astronomers thought it did, Seneca (approximately 4 BCE – 65 CE) claiming that Sirius' red color was more pronounced than that of Mars. Ptolemy compared it to Betelgeuse.

Some scientists have wondered if there might have been some local condition at Sirius that caused it to shine red for a time, perhaps a tidal interaction with the possible star Sirius C or the burning of hydrogen in a thin shell around Sirius B. There might even have been a small but dense cloud transiting across line of sight through the local interstellar medium.

Alternatively, there was never a red Sirius. This seems the most likely. Perhaps the observation was distorted by Earth's atmosphere. Or perhaps there was a cultural expectation that Sirius should be red. This is Aratus, describing the star in his work *Phaenomena* of the third century BCE: "A star that, keenest of all, blazes with a searing flame and men call Sirius. When he rises with Helius, no longer do trees deceive him by the feeble freshness of their leaves. For easily with his keen glance he pierces their ranks, and to some he gives strength but of others he blights the bark utterly."

Searing flames are red, so maybe the ancients saw what they expected to see. Sirius had already assumed a fixed place in their cultural imagination. A fixed place it has retained.

From early human history, the star acquired both canine and fiery connotations. In the Akkadian language of the ancient Near East it was 'The Dog-Star of the Sun,' a dog in the sense that it was a companion to the Sun at its strongest. Ancient philosophers took the view that dogs were by nature dry and hot and that Sirius was the bringer both of heat and of the disease of dogs' heat, rabies. A ninth-century image of the constellation shows a ferocious dog whose tongue is literally a flame. Sirius is burning.

In the Egypt of the pharaohs, Sirius had a different power associated with the annual flooding of the river Nile. The star's appearance above the eastern horizon just before the Sun seemed like a prophecy of this great event, and several temples were oriented to let the rays of the star fall upon the altars. Later, Sirius even became attached to the production of honey, which according to Pliny: "Is engendered from the air, mostly at the rising of the constellations, and more especially when Sirius is shining."

Modern poets have not forgotten Sirius and the summer. We still talk of the dog days, meaning that hot part of the year when Sirius is in the ascendant. It is a time for lying back, more than for achieving. So in Dylan Thomas's poem "I See the Boys of Summer," it is "the dogdayed pulse of love and light" that burns in the boys' throats. The poet, watching these boys wiling away their summer in love and girls and folly, predicts that they will become "men of nothing" and sons of the Sirius summer.

The Size of a Nose

The human encounter with Sirius has been about scale – the brightness of the star, its distance from Earth, its invisible companions. There is one more piece of scale in its history, a spirited young giant with a 6,000-foot-long nose.

The novel *Micromégas* (1752) is the work of the French Enlightenment thinker Voltaire (1694–1778). It belongs alongside his greater work *Candide* as a steady and at times hilarious examination of where humanity belongs in the cosmic scale. *Micromégas* is a grand satire upon the voyage of discovery. Instead of going from Earth to learn more of the universe, a visitor called Micromégas comes from Sirius to discover how small we are.

His journey is easy. He travels from Sirius by way of gravity, sunlight and convenient comets as far as Saturn, where, meeting the secretary of the Academy of Saturn, they agree to journey on together through the Solar System. They pause briefly upon Jupiter but bypass Mars "like two travelers disdainful of a bad village cabaret." Finally, they reach Earth, "a pitiful sight to those who had just left Jupiter." Everything is so small that it is only by the chance of a broken necklace that the two voyagers bend down and, with the help of a microscope, identify first a whale and then a boatload of philosophers.

Micromégas is at first impressed. He finds the humans' grasp of science and mathematics quite amazing for such small creatures. But it is when the "intelligent atoms" speak of Earth's business that Micromégas changes his mind. He learns of wars taking place over mere piles of mud. Things become less appealing still when the philosophers talk of the soul and human identity. None of them agree. One quotes some Greek because "one should always cite what one does not understand at all in the language one understands the least."

Micromégas offers one simple truth, that "many more things are possible than one would think." As this seems to the philosophers to be not quite enough,

Micromégas promises to leave them a beautiful book in which "they would see the point of everything." Micromégas does indeed leave them just such a book. It is empty.

Why is Voltaire telling this story? To establish the cosmic scale, the theme that underpins the story of this star. That which appears brightest and alone in the sky is both less bright than other stars and also not alone, but it has been the star in whose company scientists have stumbled towards the true scale of the universe. If someone from Sirius were ever to come, he might indeed have Micromégas' 6,000-foot nose.

Chapter 10

Looking for Footsteps

Observation: Craters, maria and mountains on the Moon
Significance: Varied lunar surface, origins of the Moon, Apollo missions
Science: Impacts, crater types, formation of the lunar seas

The Moon landings came to a sudden end. We have not been back since the last Apollo mission. But the relationship between Earth and the Moon is just too close to allow for the impossibility of return. We talk about it, send probes to map it and even smash things into it. They are all good science, for sure. They are also ways of keeping the connection alive.

In this second chapter devoted to the Moon, we are going to extend our observation to the details of craters, seas, mountains and walls. We will look at the history of the Moon and what it tells us about Earth. The Moon displays our planet's early history such that understanding the Moon tells us a great deal about how Earth evolved. We will then take a journey to the Moon, one not of sight but of the mind. There, in the company of one of the greatest of all astronomers, we will find both scientific insight and personal grief.

A Trilogy of Craters

The 5-day-old Moon reveals one of the treasure troves of lunar observing, the three craters Theophilus, Cyrillus and Catharina. They are easy to spot in the early evening by scanning the terminator until you reach a perfectly round, walled crater. Start from the south – binoculars preserve the instinctive compass points of naked-eye observing – and you will pass over a pitted area before coming to Theophilus,

M. Marett-Crosby, *Twenty-Five Astronomical Observations That Changed the World: And How To Make Them Yourself*, The Patrick Moore Practical Astronomy Series, DOI 10.1007/978-1-4614-6800-4_10, © Springer Science+Business Media New York 2013

which on this fifth day after the new Moon lies mostly in shadow. But the tops of its central peaks and part of its eastern wall might well be catching the first rays of the Moon's morning.

The Theophilus crater recalls the Christian saint, Theophilus, patriarch of Alexandria (d. 412). The name appears first on Riccioli's map of 1651. The crater was labeled previously as Golius and Ferdinand Franciscus in the two versions of Van Langren's map (1645 and 1648) and as Mons Mosclius by Hevelius (1647).

Over the subsequent hours and days, Theophilus gradually reveals itself as one of the great lunar craters, sharply formed and achingly clear through a telescope. To see it in full, it's best not to wait too long and to use a lunar filter so as to exclude glare. Certain features emerge. Theophilus has finely terraced crater walls, still largely intact. These walls are fantastic to watch as the shadows stretch across them. The crater floor is flat and largely undisturbed by impacts, telling us that it is relatively young. Theophilus has a broad central mountain with a peak that catches the sunrise.

Theophilus summarizes lunar history. It is about violence. The Theophilus event itself was powerful enough to raise a crater wall and central mountain. Compare it for scale with the neat little Mädler Crater, which is to the left if you are looking through a telescope without correcting apparatus. Mädler is a tidy punch to Theophilus' missile strike. To put it in the language of crater science, Mädler is a simple crater while Theophilus is complex.

There is further evidence of the power of the Theophilus impact. Nearer to full Moon, the crater's high albedo stands out bright against the dun surrounds. Furthermore, although not readily visible through amateur telescopes, the crater has a skirt, sometimes 5 km wide, of ejecta lying around it. It is suggested that the crater was formed around a billion years ago, one of the last great impact events on the lunar surface.

The violence of the lunar past is more visible as we place Theophilus in its context. Moving southwest, it is easy to observe the full trio of craters, all about 110 km in diameter, as captured in the image shown.

Fig. 10.1 The Theophilus, Cyrillus and Catharina chain of craters (Credit: Mark Crossley, http://www.wilmslowastro.com/)

Immediately above (as seen in a telescope) Theophilus lies Cyrillus, recognizably a crater, although Theophilus has smashed away the northeastern quarter of its rim. Cyrillus has a smaller impact crater, Cyrillus A, within its bounds, and there are rilles and ridges on the floor that contrast with Theophilus' 'clean' appearance.

Cyrillus and Theophilus are related by name as well as place. Cyril was Theophilus' nephew and his successor as patriarch of Alexandria (d. 444).

Beyond Cyrillus lies Catharina, the most ghostly of the trio, rugged and deformed. The crater Catharina completes the Alexandrian theme, recalling St. Catherine of Alexandria, a Christian martyr. By counting impact events, a rough-and-ready but reliable means of assessing age, it is clear that Catharina is the oldest crater of the three. One of Catharina's sub-craters is itself some 45 km wide and has laid waste to the Catharina's northern wall.

We can add more detail to our understanding of the Theophilus crater by way of the Moon Mineralogy Mapper (M3), a NASA instrument carried on the Indian *Chandrayaan 1* lunar probe during its mission from October 2008 to August 2009. M3 was an imaging spectrometer capable of examining the lunar rock types. It discovered several types of previously unknown lunar material including, at the peaks of Theophilus and in isolated patches on the floor, the magnesium-rich mineral spinel. The mineral appears in only one other place thus far identified, and that is in the Moscoviense Basin on the far side of the Moon. This evidence suggests that the Theophilus impact excavated the spinel from a source deep within the lunar crust.

Fig. 10.2 Theophilus crater, image taken on 18 August, 2009, 18 in. reflector, infinity 2-1M Camera (Credit: Wes Higgins, http://www.higginsandsons.com)

The Science of Craters

All the terrestrial planets and many satellites of the giant planets exhibit craters, the evidence of powerful impacts upon the surface caused by incoming rocky projectiles.

It was not always obvious to scientists that craters are the remains of impacts. With the early lunar maps drawn in the seventeenth century came a consensus that they were caused by volcanic processes, and the few dissenting voices – Robert Hooke was one – remained largely ignored until the work of American geologist Grove Karl (G. K.) Gilbert (1843–1918), whose 1893 article, *The Moon's Face*, presented serious arguments for impacts from measurements of depth to diameter ratios and other features, the rays and ejecta.

(Robert Hooke (1635–1703) was a British philosopher-scientist famous for his experiments with light and also an astronomer. He built his own telescopes and attempted to measure stellar parallax by observing the star Gamma Draconis. He also studied Saturn and lunar craters, proposing an impact model. Hooke was also the first observer to draw an individual lunar crater, Hipparchus, which he published in his *Micrographia* of 1665).

Many years of academic disputation passed before the impact theory was accepted, but finally the work of D. M. Barringer and Gene Shoemaker (1928–1997) demonstrated an impact origin for Meteor (Barringer) Crater in Arizona. Shoemaker applied the understanding of this crater to the Moon. It is now generally recognized that craters on all rocky Solar System surfaces are formed through an initial impact and compression, then excavation and finally modification.

Crater science makes a distinction between different crater types. The smallest are micro-craters, tiny holes caused by comet dust or other Solar System debris with considerable destructive potential for orbiting satellites. When the Hubble Space Telescope's solar arrays were exchanged in 2002 after over 8 years in orbit, the wings were found to have suffered 149 impacts, of which 62 craters were caused by man-made space debris and 52 were micrometeorite craters. The origin of 35 could not be determined.

Simple craters are bowl-shaped and without other features. Theophilus illustrates the characteristics of a complex crater – terraced walls, a central peak, and in this case rays of ejecta. Most craters are round but some are elongate. There are several of these on the Moon, including Schiller and Messier, and a vast example on Mars, the Orcus Patera. These demonstrate the effect of the angle of impact upon crater shape.

Cratering allows us to construct a broad history of the Moon divided into epochs. The dates are uncertain, and vary a little in published accounts. The earliest period between the Moon's initial formation and 3.92 billion years ago is the Pre-Nectarian, during which the lunar crust formed and bombardment was intense. Some of the major impact basins were created at this time.

The Nectarian period (3.92–3.85 billion years ago) was marked by further heavy bombardment and the formation of the basin now occupied by the Mare Nectaris. The Imbrian period (3.85–3.1 billion years ago) is identified with the Mare Imbrium and with lava flows. It is followed by the Eratosthenian period, between 3.1 and 1 billion years ago, when many of the older visible craters were formed.

Finally, the Copernican period is dated from 1 billion years ago. This is the time of younger craters such as Copernicus and Tycho.

The point is that the Moon wears its story on its skin. This is in contrast to Earth, where the movements of continental plates and the process of subduction have concealed these earlier times. In observing the Moon, we are glimpsing our own primeval history. In coming to understand our nearest rocky neighbor, we learn a great deal of what our own planet once was and would be without the effects of an atmosphere and plate tectonics.

Dry Seas

Our second set of lunar observations begins at Theophilus. The flat land alongside it is the Mare Nectaris (Sea of Nectar), a pallid plain that flows northwards into the distinct but connected Mare Fecunditatis (Sea of Fecundity) and then Mare Tranquilitatis (Sea of Tranquillity), with the Mare Crisium (Sea of Crisis) lying a little apart and furthest to the north. Theophilus and Mädler lead directly into Nectaris, a circular area about 350 km in diameter. Running a telescope across it reveals very little of anything, although as you reach its southern 'shore' it is worth spending time with crater Fracastorius, which has been flooded such that it now

Fig. 10.3 The crater Fracastorius on the edge of the Mare Nectaris, Newtonian Orion Optics 250 mm @f/6.3, Unibrain camera Fire-i785, barlow 3X, red filter (Credit: George Tarsoudis, www.lunar-captures.com)

appears like a 'bay' off the great 'sea.' This observation invites questions. What are the lunar seas, and what is the nature of the flooding that has filled Fracastorius?

There are clear differences between maria and the cratered regions. The lunar highlands stand some 2.75 km above the seas and are topographically complex, whereas the maria are very flat, with slopes of less than 1,000 m according to measurements made by the Clementine spacecraft. Maria rocks collected by Apollo and Luna missions are fine-grained volcanic basalt. It is thought that the maria are coated in frozen lava drawn from far beneath the crust as it flowed from ancient impact basins. The magma beneath the surface emerged from the wounds, flooding wide areas of the surface with this uniform basalt.

There is evidence to suggest that the last of these lava flows did not cease until around 1 billion years ago. Since that time, the Moon's crust has been too thick to allow any liquid magma, if there is any, to emerge.

It is this flooding that gives rise to a particular feature of maria, 'ghost' craters so flooded by lava that only the outline of the crater rim survives.

Craters have human names and lunar seas don't, a convention going back to the earliest lunar maps. Their current names are mostly derived from Riccioli's map of 1651. The crater Fracastorius commemorates Girolamo Fracastoro (1478–1553), an Italian polymath who proposed an alternative theory of planetary motion called homocentrism. He also wrote an epic poem that coined the name syphilis for the disease then ravaging Europe.

Returning to the telescope and moving north from Fracastorius, Mare Nectaris is separated from Mare Tranquilitatis by a rugged area, but the mare terrain is obvious once it restarts. Tranquilitatis is one of the oldest of the lunar seas broken on its eastern side by the crater Cauchy, which shines like a fallen star on the face of a full Moon. There are some interesting but faint features nearby, including the Rupes Cauchy and two tiny domes that appear like freckles under high magnification and good seeing.

If you fancy a late night lunar session with a 19- or 20-day Moon, track north from Theophilus to the western rim of Tranquilitatis to spot the tiny, lonely Maskelyne crater with Arago further into the daylight. This region has been touched by humanity reaching towards the stars, first by probes – *Ranger 6* in 1964, *Ranger 8* in 1965, *Surveyor 5* in 1967 – and then by people when, on July 20, 1969, *Apollo 11* landed some 150 km east of the Sabine Crater.

Fig. 10.4 The Apollo 11 landing site captured from just 24 km above the surface provides LRO's best look yet at humanity's first venture to another world (Credit: NASA/GSFC/Arizona State University)

It is impossible to see any evidence of human activity through amateur instruments, but NASA's Lunar Reconnaissance Orbiter (LRO) has imaged the Apollo landing sites in such detail that it is possible to see what we left behind.

Fig. 10.5 The Little West crater and the Lunar Module, photographed by Neil Armstrong (Credit: NASA)

Apollo in the History of Humanity

It is worth placing this orbiter picture alongside one taken from the ground, seeing the same stretch of the lunar surface but with the shadow of an astronaut cast over it.

This is, of course, an *Apollo 11* image, redolent with international rivalry and personal bravery, ambition and technical prowess, that took humanity beyond Earth to walk, ride and play golf on the Moon. It is perhaps the greatest story that there is to tell. But alongside Apollo's achievement, it is sad to acknowledge that we have not been back to the Moon since.

The promise of Apollo was twofold – that humanity's mission to the Moon and other worlds would continue, and that going to the Moon would change Earth.

The first part of that promise has come true only in part. Remote exploration of other worlds has flourished and to great scientific effect. If we compare our understanding of the Solar System now with that in place when Apollo left the Moon, we can point to substantive advances in our knowledge of rocky planets, gas giants and the Sun. Probes have met with comets and asteroids, and we are on our way towards the Kuiper Belt and an encounter with Pluto. But no human has entered deep space since Cernan, Harrison and Schmitt turned *Apollo 17* back to Earth in December 1972.

Does that matter? The economic costs of sending humans into deep space, whether to the Moon or beyond, are vast, and would swallow any or all other space projects. It has been argued that remote probes can do better science than humans. Probes need less looking after, and they don't have to be brought back. Set against these points are some economic counter-arguments and the fact that Apollo Moon rocks have continued to be an important resource for lunar science. We will see how they continue to matter in a later chapter. Above all, there is a power that only

Fig. 10.6 Earthrise over the lunar surface, December 1968 (Credit: NASA)

manned missions have both to inspire and to unite. We seem stubbornly reluctant to find low orbit space missions enough.

What about that second part of the promise, that going to the Moon would change Earth? This is a big question with big, demanding answers. Let's take one image as an icon of the whole.

During the *Apollo 8* mission, when Frank Borman, Jim Lovell and William Anders became the first humans to leave Earth orbit in December 1968, they not only caught this celebrated image of Earthrise but also broadcast live from their spacecraft, reading from the Book of Genesis and speaking of what they were experiencing. Lowell said: "The vast loneliness is awe-inspiring, and it makes you realize just what you have back there on Earth."

This was Apollo's message to "the good Earth," as Frank Borman called it in the same broadcast. It is paradoxical but telling that the only human venture thus far into deep space has brought back an understanding of the fragility of our planet in the vast setting of the Solar System and the stars.

Climbing Mountains

Our third lunar observation takes us back to Mare Serenitatis. The only crater of any size to interrupt it is the 100 km wide Posidonius located on its western edge, Nectarian or even older and flooded with lava. Its central peaks peer out like submarine periscopes.

A 19-day Moon is particularly good for observing Posidonius and its neighbor Le Monnier, the crater explored by the Russian *Luna 21* mission in January 1973.

A little earlier in the lunar month, it is possible to enjoy good views of lunar mountains. As Tranquilitatis slides towards the night, move away from the terminator and towards the adjacent sea, the Mare Imbrium. It's clear even under full Moon conditions that these two dark areas are separated by lighter regions to both north and south, leaving a gap in the center. By a 20 or 21 day Moon, that gap is becoming shadowed, and the lunar ranges around it are sharp. To the north and closer to the terminator at this late stage in the lunar cycle lie the Montes Caucasus, worth surveying in binoculars for their cold chemistry of shade. The same observation with different shadows can be made at around day seven after a new Moon. These mountains are 400 km or so long and rise high above the maria.

Move eastwards from the Caucasus, and there are two distinct craters, Autolycus and Aristillus, a lovely observing pair. A little to the north of Aristillus lies isolated Mons Piton with Mons Pico, a more difficult target, further in the north. At the other end of the lunar month, around day 7 once again look for the first rays of dawn as they touch these mountain peaks.

To the south and stretching towards the craters Eratosthenes and Copernicus stand the bulky Montes Apenninus. In really good seeing, it is possible to observe not only these mountains in detail but also Hadley's Rille, a sinuous channel in the surface of the Mare Imbrium. It was in the foothills of the Apennines that *Apollo 15* astronauts David Scott and Jim Irwin spent 3 days in July 1971.

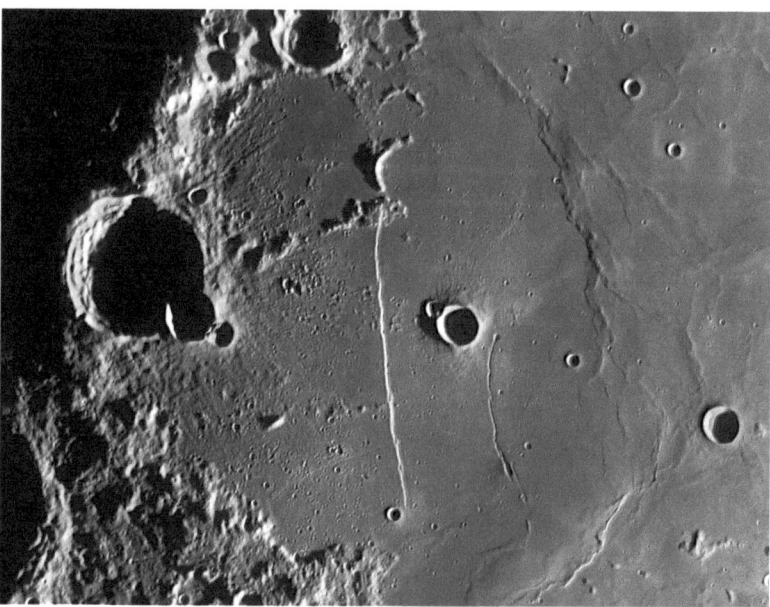

Fig. 10.7 The Rupes Recta, imaged on 24 August 2008 with a Celestron 14 at F/19 (extender QX1.6 Takahashi) and Skynyx 2.1M video camera, exposure time: 45 ms (Credit: Christian Viladrich, http://christian.viladrich.perso.neuf.fr/)

Finally, anyone observing the Moon will want to find its most celebrated fault line, the Rupes Recta or Straight Wall. On the face of a 9-day Moon, identify the two craters Copernicus and Tycho. A straight line between these two features takes you very close to the Rupes Recta on the sunlit side, just as the Mare Nubium touches highland terrain to its south. The Straight Wall is 120 km long and around 250 m tall. The distinctive, almost geometric nature of its shadow will vary enormously depending on the angle of illumination.

Where Does the Moon Come From?

Mountains, rilles and walls – what part do they play in lunar history? The origins of the Moon are shrouded in uncertainty, and different theories have enjoyed support at different times.

Broadly there are four ideas as to where the Moon came from. The first is capture, suggesting that the Moon, like the Martian satellite Deimos, is essentially a foreign object drawn in by Earth's gravity and unrelated to the geology of Earth. The second is its opposite. The co-accretion model proposes that the Moon formed from a debris cloud in orbit around the young Earth. A third theory, fission, argues for the Moon as a chunk of material spun off from Earth, while the fourth and most violent theory proposes that the Moon was caused by a giant impact between Earth and another planetary embryo. This Big Whack origin (more formally, the giant-impact hypothesis) is currently the most favored.

Seismic activity indicators left on the surface by the astronauts suggest that the Moon has a nickel-iron core much like that of Earth but smaller. This indicates that the Moon's origin is different to that of other rocky bodies.

Meanwhile, analyses of Moon rocks reveal that Earth and the Moon share much basic geochemistry but not all. The Moon is notably lacking in some elements. This implies that the impact that formed the Moon occurred at a point in Earth's history when some differentiation of material had already occurred and certain heavy elements had sunk into Earth's crust. Molten rock is thought to have covered the early Moon – some of this remains as anorthosite in the lunar highlands. Other materials sank, and the little world began to cool, but there was clearly still enough liquid magma to rise into the early impact basins. There are no obvious lunar volcanoes. Lunar rilles are thought to be collapsed tubes where lava once flowed, and there is some evidence of ancient leaking from the interior in the ejecta around what are known as dark halo craters such as those inside the crater Alphonsus.

Apollo Moon rocks indicate that the Moon retained its magnetic field until some 4.2 billion years ago. This is longer than it should have endured. The Moon has no global magnetic field now, yet it displays magnetic anomalies in certain places. The causes of this enduring magnetism are still being debated.

The Moon has no atmosphere. This was perceived by Roger Boscovich (1711–1787), a Jesuit priest and polymath who in 1753 argued that the instant disappearance of stars as they were occulted by the Moon demonstrated that there was nothing above the surface except the vacuum of space. Scientists now talk of a lunar exosphere, a tenuous gathering of about 100 molecules per cubic centimeter, compared to about 100 billion billion molecules in the atmosphere around Earth. Most of these molecules, arriving in the solar wind or caused by meteoroid impacts and outgassing from the interior, seem to fall towards the surface during the lunar night, creating the 'horizon glow' noted by the *Surveyor 7* lander in 1968. This immensely fragile environment will be the focus of a NASA mission, the Lunar Atmosphere and Dust Environment Explorer (LADEE), launched in 2013.

What About My Writing 'A City of the Moon?'

So wrote Johannes Kepler in a letter of 1623. He was describing his response to an observation he had made some 14 years before, "something wonderful and remarkable, towns with round walls, as one can see by the shadows." He was describing, surely, lunar craters like those we have been visiting in this chapter. The sight inspired him "to leave this Earth and go to the Moon."

The result was a journey of the mind that Kepler named his *Somnium*, "The Dream," which appeared in print only after he died in 1630. It was little read for many centuries, which is a pity, for in Kepler's *Somnium* we have the missing link between the flighty dream narratives of the ancient and medieval world and recognizable science fiction, that is, fiction with science at its center.

The core of the story is a trip to the Moon undertaken by a character called Duracotus. Kepler describes how Duracotus first studied under Tycho, thereby ensuring a proper foundation in contemporary astronomy. But Duracotus also receives from his mother knowledge of a spirit that can lead him on this journey, arduous and beset with "terrific cold and difficulty in breathing." But it is worth it once he reaches the Moon, because Duracotus then receives "such a different view of the motions and sizes of the planets that it must surely have a wholly different system of astronomy."

It does, and much of Kepler's *Somnium* is concerned with that astronomy, with the geography of the Moon as Kepler himself observed it. This story is a fantasy but is grounded in observations, a first mixing of physics with literary imagination. The text itself is outweighed by a factor of 8 to 1 by the 223 footnotes that Kepler added to explain every detail. It is here, among the footnotes, that Kepler writes: "The purpose of my dream is to use the example of the Moon to build up an argument in favor of the motion of the Earth."

This is science fiction with a serious purpose. Kepler is trying to imagine how the motion of Earth would appear from the Moon, the ultimate proof of a Copernican Solar System. He even conceives of how Earth might appear from the Moon, anticipating by some four and a half centuries the experience of the Apollo astronauts. He writes with a poetic flourish of how: "The eastern part is what looks like the front of a human head cut off at the shoulders, approaching a young girl with a long dress to kiss her. She, with hand extended backwards, is enticing a cat."

Kepler justifies such imagery because it is a "refutation of the argument, based on sense perception, against the motion of the Earth." The stretching hand, by the way, is Scandinavia and Great Britain. Squint at a map, and it is almost possible to see what he means.

The *Somnium* is a piquant, clever book. Kepler uses the language of spirits and dreams in order to present an argument in favor of new knowledge. "You observe sound reasoning," he tells us in one of his last footnotes. But it is also a very personal text, and it captures a grief in Kepler's life. Duracotus is inspired in all he does by his mother. She, too, used to commune constantly with the Moon. It is Duracotus' mother who has access to magic herbs and the spirit who takes the astronomer up to the Moon. Outside the pages of fiction, Kepler's own mother Katharina was tried for witchcraft. His aunt had been burned at the stake for the same offence. After her arrest in 1615, Kepler used all the resources he could to secure his mother's acquittal, but in August 1620 she was taken to prison and shown the instruments of torture. In the end, there was no trial, the evidence against her mere gossip and calumny.

Kepler, meanwhile, had used his fiction to describe a mother who "delighted in the knowledge I had acquired about the sky." Perhaps he was trying to preserve his memory of his mother against the lies. As a young boy, Katharina had shown him a comet and the Moon. It was enough magic to inspire a whole lifetime.

Chapter 11

A Whispering River

Observation: The constellation Eridanus
Significance: The search for extraterrestrial intelligence
Science: The Drake equation, SETI

Our eleventh observation deepens our acquaintance with the variety of stars. It also extends our knowledge of galaxies, taking us to some of the tantalizing nebulae that lie along the banks of Eridanus, the constellation of the River. Eridanus wanders from beneath Orion's feet deep into southern skies. It provides some interesting observing challenges and has been the setting for some extraordinary science. It was in this constellation that we thought to hear the voice of intelligent life beyond our planet.

The Young River

Eridanus is a rare constellation, representing a natural but not a living thing. It meanders from celestial north to south and flows out among constellations familiar to southern hemisphere observers – Hydrus, Phoenix, Horologium the Clock. Its source lies beside Rigel, the bright, right-side foot of Orion.

Eridanus' first star is inconspicuous Lambda Eridani, almost directly beneath Rigel. The bright star a thumb's width away from Rigel is Beta Eridani, the star Cursa. Its name is derived from an Arabic homage to Orion, meaning "Chair of the Central One."

M. Marett-Crosby, *Twenty-Five Astronomical Observations That Changed the World: And How To Make Them Yourself*, The Patrick Moore Practical Astronomy Series, DOI 10.1007/978-1-4614-6800-4_11, © Springer Science+Business Media New York 2013

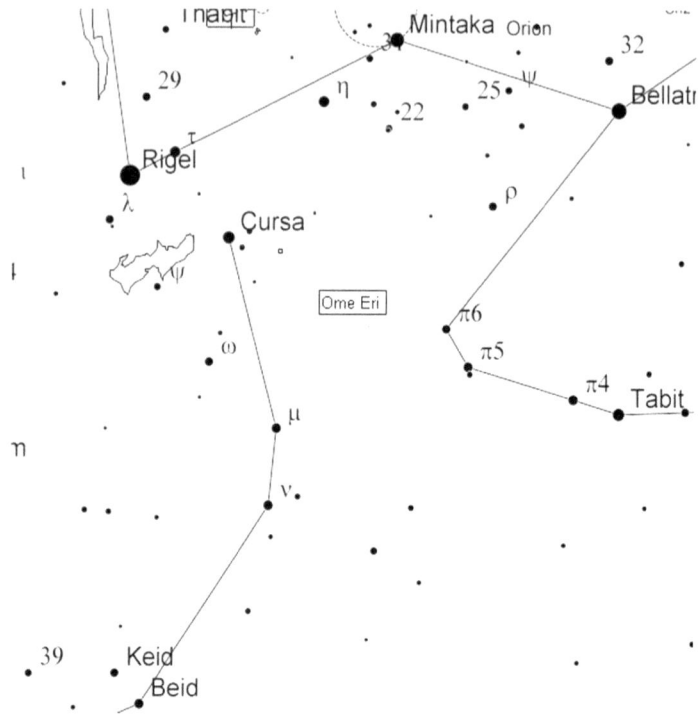

Fig. 11.1 The northern part of Eridanus flowing from the feet of Orion (Cartes du Ciel)

Cursa is an A-type star, like Sirius, some 89 light years away and probably some 300 million years old. Its age is important, because it excludes Cursa from membership of the Ursa Major Moving Group mentioned earlier. Richard Proctor (1837–1888) was the first to identify a "community of motion" among the stars in this area of the sky, and Cursa was regarded for some time as an outrider of the group. However, modern studies suggest it is too old, already in its dying days.

Orion was provided with two chairs in early Arabic interpretations of the sky. Together with Lambda Eridani and two other stars, Cursa marked the hunter's 'foremost footstool,' while the stars of the constellation Lepus served as his 'hindmost footstool.'

Cursa is a useful anchor for the river constellation. From Beta, Eridanus swings about before settling, more or less, onto a southward course. To meet up with Eridanus again, though, it's easiest to return to Rigel and identify Lambda Eridani beneath it. These two point together southwards across a fairly bland stretch of sky towards a gathering of stars all of about the same brightness. It's tempting here to wander too far down towards the more obvious bright star beneath. Move back up and identify the little group. It's worth it.

You'll find here a pair of stars that look a lot like the twins of Gemini. They mark the bend in the river where Eridanus turns south. A small telescope will separate these stars. The upper star of the pair is Omicron 1 Eridani, whose Arabic name

Beid is evocative of a forgotten asterism, the Ostrich's Nest, Beid referring to the ostrich's clutch of eggs. The star is a variable of the Delta Scuti type, with small variations in magnitude and multiple periods of brightness caused by its rich helium atmosphere growing opaque as the gas becomes ionized.

The real prize in this part of Eridanus, however, is Beid's twin star Keid, the eggshell star, also labeled Omicron 2 Eridani but known to astronomers as 40 Eridani. Only 16.4 light years from Earth, this is one of the closest stars to us despite its inconspicuous appearance. It is not a single or double star but a triple system.

The main star of the system is classified as a flare star, a type of variable that can explode into sudden and brief brightness before dimming again. The nearest flare star to Earth, some 8.7 light years away, is UV Ceti (also referred to as Luyten 726–8 or Gliese 65). UV Ceti is a double star of which both components are flare stars, although one, UV Ceti A, is much less variable than the other. UV Ceti B varies widely in magnitude, with flares blossoming upon it every 10 h or so. These relapse after just 10 or so minutes. Sadly, UV Ceti is not easy to observe without very good astronomical equipment and even better patience. To undertake this kind of variable star observing, you can consult the excellent observing section in Martin Mobberley's *Cataclysmic Cosmic Events and How to Observe Them* (Springer, 2008).

40 Eridani is not alone. With reasonable magnification, a telescope will split the principal star from 40 Eridani B, an accessible white dwarf star. 40 Eridani B has a magnitude of +9.5 and is burning at some 16,700 K. It is joined in the system by the red dwarf 40 Eridani C, which at magnitude of +11.2 is outside our reach.

Where Does Spock Live?

The 40 Eridani triple system has found its way into science fiction. In Frank Herbert's *Dune* universe, Eridani A is the star of the planet Richese, whose ruling lords are able to create an artificial moon. But it is also, possibly, the star around which orbits Vulcan, home of Dr. Spock of "Star Trek." Although not mentioned in the television series, the literature around "Star Trek" suggests that either Eridani A or Epsilon Eridani was the star upon which the young Spock gazed. "Star Trek's" creator Gene Roddenberry confirmed his own preference for 40 Eridani, writing that: "Based on the history of life on Earth, life on any planet around Epsilon Eridani would not have had time to evolve beyond the level of bacteria. On the other hand, an intelligent civilization could have evolved over the aeons on a planet circling 40 Eridani. So the latter is the more likely Vulcan sun."

The Boat and Its Galaxies

After observations among the omicrons, it is a relief to move southwards to Gamma Eridani, one of few stars in the constellation that is visible without a telescope. It rests in a fairly empty part of the sky beneath 40 Eridani and is the boat on the river, according to its Arabic name of Zaurak. It is a pretty red-orange star of the M or Betelgeuse spectral type, one of the widest classes of stars and with the most complex of all stellar spectra.

Zaurak is a useful guide star for the next observation. Keeping it towards the bottom of a binocular field of view, you will pick up on a triangle asterism towards the top right of what you're seeing. These are, in increasing distance from Zaurak, first Pi and then Delta and finally Epsilon Eridani. The line between Delta and Epsilon also points further into the constellation at the stars Zeta and Eta Eridani.

Epsilon is not much to look at. Although one of the closest stars to us, 10½ light years away, it has a magnitude of +3.7 and shines the deep orange color that is characteristic of K-type stars. It is a variable star of the DY Draconis variety and in the year 2000 was confirmed to have a planet in its orbit. Circling at 3.4 AU around the star, it is one of the many 'hot Jupiters' that have been identified in exoplanet studies.

The K spectral type is easy to observe. Giant K stars are common and include Pollux in Gemini, Dubhe in the Big Dipper, Arcturus in Boötes and Aldebaran in Taurus. 61 Cygni is a visible K dwarf and the first star to be measured by parallax.

More startling even than the planet, astronomers have detected a dust cloud in Epsilon Eridani's orbit. This cloud was imaged in unprecedented detail through the James Clark Maxwell telescope on Mauna Kea, Hawaii, in 1998.

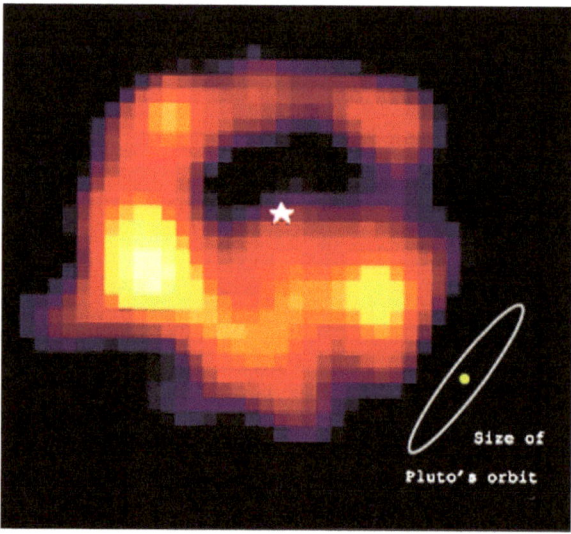

Fig. 11.2 Submillimetre wavelength view of a ring of dust particles around Epsilon Eridani, taken with the SCUBA camera at the James Clerk Maxwell Telescope. The false-colour scale is brightest where there is more dust (Credit: JCMT/JAC/http://outreach.jach.hawaii.edu)

What is this image showing us? Epsilon Eridani is marked in the center, and the bright areas around it indicate concentrations of dust at a distance of about 65 AU from the star. We are looking at something analogous to the Kuiper Belt in our own Solar System. It has been suggested that some of this dust is being drawn in towards the star to form a pair of 'warm belts' at around 3 and 20 AUs. It is expected to have comets much like our own. An additional planet has also been proposed, but has not been confirmed.

Epsilon Eridani could have been the scene for an even greater breakthrough. It nearly changed us all forever. In April 1960, a 2-month experiment began at the National Radio Astronomy Observatory at Green Bank, West Virginia, called Project Ozma. Led by Dr. Frank Drake, this was the first attempt to use the fledgling science of radio astronomy to listen for evidence of alien technologies. Two nearby stars were chosen, Tau Ceti in the Constellation of the Whale and our own Epsilon Eridani. The listeners tuned in to a band near the 21 cm line of hydrogen. They reached out to hear, and, indeed, they heard. Something was speaking to them. This is how Frank Drake describes the moment in his book *Is Anyone Out There?*:

> We set up the recorders again and readied ourselves for the long wait. But scarcely five minutes passed before the whole system erupted. WHAM! A burst of noise shot out of the loudspeaker, the chart recorder started banging off the scale, and we were all jumping at once, wild with excitement.

However, our view of the universe remained unshattered. There was indeed a signal from an object in the sky, but it was a passing airplane. Epsilon Eridani was silent.

It will continue to attract our attention, though. It is already a well studied star, and may be approaching a close encounter with another stellar neighbor, the star UV Ceti. Recent research has suggested that they will pass within a light year of each other in about 31,500 years or so, close enough for UV Ceti to pierce Epsilon Eridani's solar system.

Returning to Gamma Eridani enables us to probe into the constellation's deeper skies, rich in galaxy observations. Although they can be tricky to locate, they reward the effort.

Draw a line in the sky between Gamma and Omicron 1 Eridani. If you imagine this line as the hypotenuse of a right angled triangle, then the star that makes the triangle is not a star at all but the planetary nebula NGC1535, labeled poetically as Cleopatra's Eye. This compact nebula really rewards the highest magnification you can attach to your telescope. With averted vision you can glimpse this ninth magnitude object and perhaps distinguish its beautiful blue disc. The central star, though, is always difficult to see without good photographic equipment to allow for long exposure.

Fig. 11.3 L/CMY image of planetary nebula NGC1535 in the constellation Eridanus made from images taken on 25.11.2000 with CB245 and 32″ f4 Newtonian from Danciger, Texas. Six 15-s subexposures in cyan, five 15-s subexposures in magenta, and five 15-s subexposures in yellow were combined with six 15-second subexposures in white (unfiltered) (Credit: Al Kelly, www.utahskies.org)

The next galaxy is harder to find, but Gamma Eridani offers a good starting place. Move down from the star until you come across a shape of four stars in a wedge. The upper pair of stars in this wedge point towards Tau4 Eridani, the guide star for two galaxies NGC1232 and NGC1300. NGC1232 forms a triangle with Tau4 and Tau3 on its longest side.

NGC1232 is mesmerizing. Millions of bright stars have clustered along its arms. It has a satellite galaxy as well, NGC1232a. The bond between these galaxies places NCG1232 in the rare company of the strange and unusual galaxies cataloged by Halton Arp. For this reason, you will sometimes find it referenced as Arp41.

Fig. 11.4 NGC 1232, image obtained on September 21, 1998, based on three exposures in *ultra-violet*, *blue* and *red light*, respectively (Credit: ESO)

Our final deep sky observation in this region of Eridanus is NGC1300, which lies above Tau4 Eridani in an otherwise depopulated part of the sky. To move from NGCs 1230 to 1300 is to transfer from father to son, Sir William Herschel having first identified the former in the year 1784, his son Sir John (1792–1871) the latter in 1835. NGC1300 is another beautiful spiral galaxy and shows a so-called 'grand design' core to the galaxy, a spiral within a spiral 89,000,000 light years into deep space.

The *Atlas of Peculiar Galaxies* (1966), the work of Halton 'Chip' Arp (b.1927), is a list of 338 bizarrely shaped galaxies and galaxy associations. Arp's achievement was to identify the dynamic rather than the static nature of galaxies, especially as they interact with each other. Our NGC 1232 and 1232A are good examples of just this sort of galactic drama.

The Mouth of the River

Eridanus flows from the dominant constellation of the northern winter into southern latitudes. Alpha Eridani is the eighth brightest star in the whole night sky and visible all year round south of latitude 33. Its name Achernar means "end of the River." In a beautiful Chinese asterism, it belongs to Crooked Running Water.

All these connections with cool water belie the star's fury. Achernar is a powerful B-type star spinning so fast that it has visibly flattened at its poles. Observations made with the Very Large Telescope in 2002 revealed a 'pronounced asymmetry.' It is the most misshapen star we know. This image shows views based on the VLT data.

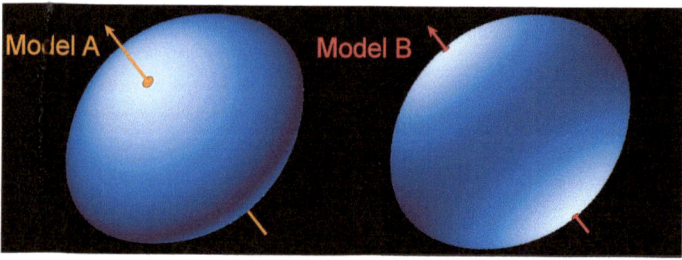

Fig. 11.5 Model view of Achernar, based on the profile measured with the VLTI. Two different models are shown: in "*A*", the polar axis is inclined 50° to the line-of-sight; in "*B*", this angle is 90° (Credit: ESO)

Achernar challenges some aspects of our theory of stellar integrity. Meanwhile, an absence over much of Eridanus questions our understanding of the universe. This constellation hosts a Cool Spot, a measurable chill in the temperature of the cosmic microwave background (CMB) identified in 2005 over a large swathe of the constellation.

The CMB is the glow, almost exactly the same in all directions, that pervades the observable universe. It is not associated with any particular star, galaxy or supernova, but rather it is the radiation still remaining from an early stage in the development of the universe. It has been extensively studied by satellites, including WMAP and Planck.

The CMB should be constant, about 2.7 K, but NASA's Wilkinson Microwave Anisotropy probe has detected a drop of about 70 micro-Kelvin (70 μK) in this area. Not much, in one sense, but enough to pose difficult questions for cosmologists, with some even suggesting that the Cool Spot is a hint of a universe beyond our own.

Filling the Void

It seems fitting that Eridanus should have been chosen as part of the first experiment to listen into the void. Project Ozma and the experience of Epsilon Eridani set the rules for the science of searching for extraterrestrial technology, a work that has been championed by SETI, the Search for Extra-Terrestrial Intelligence.

Much of this search is focused upon nearby, Sun-like stars. The 21-cm band near hydrogen is still the spectral region of choice. First in Project Cyclops and then in the NASA SETI program of 1992 these hunts were expanded. The task is huge. If the goal is to achieve a SETI success within a century, then the mathematics of detection suggest that we have very little time to hear anything that is being sent to us from any one direction, perhaps just a second or less per star.

Given this, should we be broadcasting our existence ourselves? Aside from any accidental clutter, we did this in the Arecibo Blast of 1974. But are these merely screams into the darkness? Is there any scientific reason for believing there is life elsewhere than on Earth with the technological capacity to receive anything we communicate?

The Drake equation seeks to answer that question. It offers a mathematics of ET. This is the classic Drake formula, with N as the number of communicating civilizations in existence now:

$$N = Rf(p) \, n(e) \, f(l) \, f(i) \, f(c) \, L$$

The right-hand side of the equation begins with the physical. R is the rate of star formation and f(p) is the fraction of those stars with planets in their orbits. We are getting to a stage in our understanding of exoplanets where that value can be determined. The next element n(e) is a measurement of the number of planets suitable for life – we will learn more about this in a later chapter – while f(l) is the fraction of these suitable planets where life has actually developed.

We approach some much more speculative numbers with f(i), the fraction of life-bearing planets where that life has evolved to an intelligent state. The next value, f(c), refers to the fraction of such intelligent cultures that develop the technology of radio communications of the sort we can detect and the desire to speak. Finally L represents the time that such cultures survive in communicative mode before being destroyed or moving on to other things.

Some important modifications to Drake's model have been proposed. These include adding measurements to indicate the fraction of planets with stabilizing moons and the fraction of solar systems with Jupiter-like planets to shield the inner, rocky ones from large impact events.

Many of these values are difficult to assess. The intelligence measure f(i) is regarded by many as the crucial bottleneck, it being the case on Earth that most life forms do not reach the level of intelligence necessary for technological development. So in the absence of numbers, we simply listen. Writing in 1992, Frank Drake stated: "The silence we have heard so far is not in any way significant. We still have not looked long enough or hard enough."

40 Eridani did not change the cosmos. Some other star might. What if SETI or other such programs were to succeed? The ramifications of discovering not only extra-terrestrial life but extra-terrestrial intelligence would be profound. Science fiction divides itself into stories of hope and stories of fear, reflecting perhaps how the human population would respond to not being alone and also no longer at the peak of evolution. The effects on religion, philosophy and our self-belief are impossible to calculate.

No Oak Tree

Eridanus has taken us into southern skies. A glance at the list of the constellations found there reveals an interesting difference. They are in general less mythological and more practical, including Telescopium the Telescope, Microscopium the Microscope and Antlia the Air-Pump, among others.

The mapping of these constellations was the work of Nicholas de Lacaille (1713–1762), a clergyman and French Academician who in 1750 led an astronomical expedition to the Cape of Good Hope. Forced to remain by unfavorable winds, he set up his observatory on the shore of Table Bay in Cape Town and, working with an assistant and a dog called Gris-Gris, set about redrawing the inadequate maps of the southern sky. It was Lacaille who divided the huge constellation of Argo Navis the Ship to create new nautical constellations such as Puppis the Stern (not the Puppy, as some have translated it), Carina the Keel and Vela the Sail. He introduced new constellations with names drawn from natural sciences.

He also extracted an oak tree from the sky. It had been placed there by Edmund Halley in 1678. The constellation Robor Carolinum, King Charles' Oak, was an attempt to honor Britain's King Charles II in the heavens by recalling the tree in which he had hidden after the Battle of Worcester in 1651. Clearly not appealing to a French clergyman, it was expunged by Lacaille.

Not content with ridding the sky of overmuch Britishness, Lacaille published a catalog of deep sky objects 16 years before Messier's, many of which were new to European astronomers. His telescope being only half an inch in diameter, his objects are easily within the reach of modern binoculars. The journey he charts through clusters such as Theta Carinae reminds us that humanity will always want to learn more, whatever is out there.

Chapter 12

The Lights of Kings

Observation: Comets and meteor showers
Significance: Gravity and Isaac Newton
Science: Comet families and reservoirs, the Oort Cloud, comets and life

Comets have fascinated us since we first looked into the sky, the cause of fear and also questions. What are they? Where do they come from? What do they mean? We have some answers, but we are still learning.

> He sat up so often long in the year 1664 to observe a comet...he found himself much disordered and learned from then on to go to bed betimes.

Take comfort from the fact that even Isaac Newton found comet-spotting compulsive. He was 22 years old when he learned the astronomer's lesson – you can't stay out all night every night. But throughout history it's been the same. There are records of crowds in the streets during great comet appearances of the past. Uncounted numbers of astronomers pointed their telescopes at Halley's Comet when it last passed us in February 1986.

Observing Comets

Naked-eye comets are wonderful events. Sadly, most comets are, most of the time, faint objects. There's good reason for this. Comets are not single points of light; their brightness is spread over a large area, and they are very subject to the effects of other bright celestial objects. Comet hunters like moonless nights a lot. It's also the case that magnitude readings for comets are misleading – a +10 deep sky object

M. Marett-Crosby, *Twenty-Five Astronomical Observations That Changed the World: And How To Make Them Yourself*, The Patrick Moore Practical Astronomy Series, DOI 10.1007/978-1-4614-6800-4_12, © Springer Science+Business Media New York 2013

like a galaxy might well be within reach of your telescope, but a comet of magnitude +10 will be more difficult to see.

The identification of comets begins with an initial letter C, P, D or X, indicating whether the comet is normal, periodic, destroyed or uncertain. P for periodic is preceded by a number. After a /sign comes the year and month of its discovery and then its name, if it has one. So Halley's Comet is formally 1P/Halley, indicating that it is periodic comet number 1 in the list, with no date of discovery, as we have known about it for so long. C/2003 R5 is a comet discovered in August 2005.

Comets with well-known paths across the sky are advertised in the astronomical press and online. These sources provide tables listing the position of the comet at given times. The tables often carry the tongue-twisting name of Ephemerides, from the Greek word for a diary or calendar. The details can then be applied to a star map.

Or the comet might be obvious. The brightest comet of the last 100 years was spotted in October 1965 by two amateur comet hunters and bears their name, C/1965 S1 Ikeya-Seki. Passing just 450,000 km above the Sun, it achieved a magnificent −10 magnitude (compare that to Sirius at 1.46). It was clearly visible in daylight.

Fig. 12.1 Comet Ikeya-Seki (Credit: NASA)

A bright comet reveals its main features – the central condensation and coma around it and the two tails, one of dust and the other of gas. The tails point away from the Sun, but there can be an anti-tail pointing in the opposite direction caused by particles moving in front of the main comet's body. When a comet of really impressive magnitude crosses the sky, it might be possible to observe other details, such as jets.

Look out also for colors. Fainter comets often reveal these, too, if you can filter out background light while not distorting the spectrum too much. Specialized comet filters claim to improve contrast in the head of the comet and to help in revealing delicate comet tails. You might then see the creamy color of the dust tail and the pale bluish gas.

Most comets, however, don't reveal themselves readily. But there is lots of help available online via the International Comet Quarterly and other websites, some of which are listed here:

- http://www.icq.eps.harvard.edu/cometobs.html offers guidance how to report a comet.
- For observers in the UK, see the BAA & Society for Popular Astronomy Comet Section at http://www.ast.cam.ac.uk/~jds/.
- The Skyhound website at www.cometchasing.skyhound.com has a news section and a monthly synopsis of what's available.
- See also http://www.nightskyhunter.com/Comets%20Visible%20Tonight!.html and the Minor Planets Center at http://minorplanetcenter.net/iau/mpc.html, with the ephemerides tables under the comets tab.

Fig. 12.2 Comet Lulin glides past Regulus the night of 27th Feb 2009 (Credit: Eric Walker)

Once you've found a comet, there are lots of things to look for and record. A good place to start is with the comet's position against the background sky and how it fits among the pattern of the constellations. Many online resources will guide you to comets this way. It's worthwhile measuring magnitude. The simplest way is by making a comparison with other stars nearby.

An interesting comet project is to assess changes in brightness between the edge of the comet's coma and its heart. Known as the degree of condensation (DC), there is a 9-point scale to score what you see. You'll need high magnification for this observation, but you'll be able to judge how condensed the coma is if you secure the comet first in the crosshairs of the finderscope.

The degree of condensation scale in simplified form looks like this:

0 = Uniform brightness
1 = Slight brightening towards center
2 = Definite brightening towards center
3 = Center of coma much brighter than edges, though still diffuse
4 = Diffuse condensation at center of coma
5 = Moderately condensed – a diffuse spot at center of coma
6 = A bright diffuse spot at center of coma
7 = Condensation appears like a star that cannot be focused – described as strongly condensed
8 = Coma virtually invisible
9 = Stellar or disk-like in appearance.

If it is comet tails you really want to see, be prepared for disappointment. Many comets appear like fuzzy balls. It's only with long-exposure photography that it is usually possible to achieve great images of comet tails.

However, you might be lucky. In the last 50 years, there have been four occasions when more than one naked-eye comet has been visible at the same time, if not from the same place. We can hope for at least one naked-eye comet in each decade. The golden rule is to expect the unexpected. With comets, nobody really knows what will happen next.

Comet Rain

Meteor showers are much more predictable. They are regular observing events, especially in the latter half of Earth's year. Meteor watching is always advertised in the astronomical press, and if conditions are right they make a spectacular sight.

Meteor showers are named after the constellations in which they seem to appear. It gives them a magical feel – we talk of the Perseids, Draconids, Leonids, Taurids and Boötids. But the connection between a meteor shower and a constellation is accidental. Meteor showers are really comet rain. They occur when Earth's orbit intersects with that of a comet. What we are seeing is the dust left behind in a comet's track passing at high speed through our planet's atmosphere. Other Solar System bodies with atmospheres experience the same phenomenon.

Thus the Draconids, which appear near the top of the constellation Draco in October, are the residue of Comet 21P/Giacobini-Zinner, while the Orionids, a meteor shower later in the same month, are leftovers of 1P/Halley. One of the best meteor shower shows is provided between mid-October and mid-November by the Taurids from Comet 2P/Encke near the Pleiades cluster.

A summer feast of meteorites is provided in mid-August by the Perseids, caused by the remnants of Comet 109P/Swift-Tuttle's passage across the sky. The image shown is a collage of Perseids taken throughout the evenings and early mornings of August 10–15, 2011.

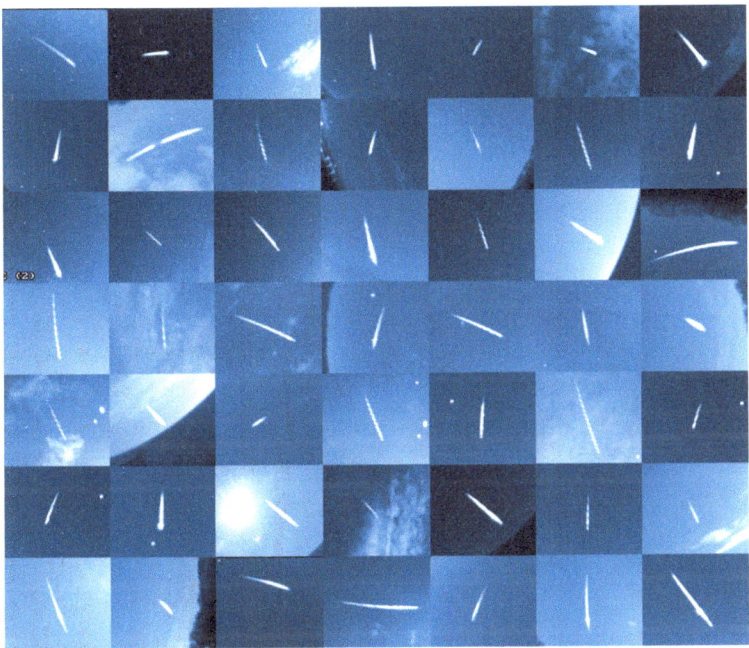

Fig. 12.3 Collage of Perseids taken throughout the evening and early morning of August 10–15, 2011 (Credit: NASA/MSFC/Meteoroid Environment Office)

People have been watching these meteor streams for a long time, but it is still possible to do some interesting science with them by counting their zenithal hourly rate. There's useful detail on how to do this at the website of the International Meteor Organization, www.imo.net, which also contains a full list of meteor showers and maps to guide observing.

Close Encounters

Comets are classified into those with orbits lasting less than 200 years (short-period comets such as 1P/Halley) and the majority, some 84.7 %, with longer orbits (long-period comets). Of the long-period comets thus far identified, most are sungrazers.

All comets are drawn towards the Sun, but sungrazers pass very close to the solar surface. They are observed from probes orbiting the Sun, and the images returned by SOHO and SDO provide dramatic images and even video footage of their close encounters.

Fig. 12.4 A sungrazer comet imaged on July 6, 2011 by the Solar Dynamics Observatory (Credit: NASA/SDO)

On July 6, 2011, the SDO observed for the first time a comet flying too near the Sun. It belonged to the Kreutz group of comets, all of which are thought to be fragments of a much larger original comet that broke up, perhaps around 371 BCE. The probe was able to record the comet as it evaporated in the solar corona less than 62,000 km from the Sun's surface. The death of this comet provided an opportunity to estimate a comet's size and mass in a new way. It is likely that, as more sungrazers are observed, we will learn more about their origins.

Short period comets are also divided into separate families. The distinctions are derived from an elegant piece of celestial mechanics, the Tisserand parameter, named after the French astronomer and mathematician Francois Felix Tisserand (1845–1896). A lifelong student of the three-body problem, his equation divides short period comets into five families by relating the orbits of Jupiter, the Sun and the comet itself. It is a measure of Tisserand's mathematical achievement that he has not only a lunar crater but also a street in Paris bearing his name.

The relationship between short-period comets and Jupiter was demonstrated very clearly when Comet Shoemaker-Levy 9 impacted with the giant planet in July 1994. It had broken into fragments some 2 years earlier. The impacts released huge amounts of heat, which revealed themselves as dark marks on the Jovian atmosphere. Further comet strikes on the planet were spotted by amateur astronomers in 2009 and 2010.

What Is a Comet?

There seem to be two reservoirs of comet cores. Some emerge out of the Kuiper Belt some 30–50 AU from the Sun, where icy-rocky fragments can be disturbed from their orbits by the gravitational effect of Neptune, causing them to start shooting towards the Sun on short-period orbits.

Their long-period cousins, similarly lumps of rock and ice, come from much further away. One possibility is that they emerge from the Oort Cloud (sometimes called the Öpik-Oort Cloud), a ring of icy bodies tens of thousands of astronomical units from the Sun. It is suggested that these objects become disturbed by the influence of other stars. They then begin their long journey inbound towards the Sun.

The Oort Cloud has not been observed, but its existence was postulated by Dutch astronomer Jan Oort (1900–1992) in 1950 and also by Ernst Öpik (1893–1985) in 1932.

There is another possible source of long-period comets even more remote than this. Computer simulations suggest that as much as two-thirds of the population of the Oort Cloud may have been formed around other stars. This would explain the composition of comets such as 96P/Machholz 1. During the comet's 2007 flyby, measurements from the Lowell Observatory revealed a lack of carbon and compounds. The scientists reported that "no other comet within our restricted database has a chemical composition even remotely similar to that of Comet 96P/Machholz 1," implying that it might be an interstellar interloper. If so, then some comets might provide clues to impossibly distant regions of space.

As they approach the Sun, comet ices begin to sublime (passing from a solid state directly to gas), forming a coma around the nucleus many times larger than the core, and giving off both gas and dust to form the tails. It is therefore the case that comets must fade over time, although this will depend on the rate of ice depletion and the chances of the comet remaining whole. As we have seen with Shoemaker-Levy 9, the gravity of Jupiter can cause catastrophic changes of orbit and the break-up of the original object.

We have come to know comets close up, sending space probes into encounters close to their cores. It takes little imagination to grasp how difficult this is. As targets for spacecraft, comets are both fleeting and hazardous. It's in their nature to move quickly and to spew out dust while doing so.

Comets, though, have the power to inspire, and none more than 1P/Halley. Observed by astronomers since at least 240 BCE, no less than five spacecraft – two Vega craft from Russia, two Japanese missions Sakigake and Suisei and the

European Space Agency's Giotto – were dispatched to meet it when it came within reach in 1986. Halley's Armada, as it became known, consisted of long-range and close observers. The Russian *Vega 1* captured the first images on March 4, 1986, but it was those taken by Giotto that inspired the astronomical world.

Fig. 12.5 Giotto's encounter with the nucleus of Comet Halley, an image taken by the Halley Multicolor Camera (HMC) on board ESA's Giotto spacecraft (Credit: Max-Planck-Institut für Sonnensystemforschung, Lindau/Harz, Germany; image supplied by Dr HU Keller. Copyright MPS)

Here was Halley as never seen before, from within 600 km of its nucleus. It gave scientists the opportunity to analyze for the first time the peanut-shaped core. They identified a 'finger,' a 'crater' and a 'chain of hills,' as well as jet sources and a central 'mountain' of some sort.

Giotto did not emerge unscathed. A single comet particle weighing one gram was enough to send the probe into a spin, and one camera was destroyed by dust. But Giotto survived, and, after a 4-year siesta, it was woken successfully from hibernation, the first spacecraft to achieve this, and dispatched to meet comet 26P/Grigg-Skjellerup. Giotto passed within 200 km of this comet's nucleus in July 1992.

The two comets made an interesting pair. Grigg-Skjellerup was both smaller and older than Halley. Most of its icy material had long since escaped into deep space. Grigg-Skjellerup was also one of few comets traveling in the same direction as the planets around the Sun. But, despite its age, there were particles some 400,000 km from its nucleus. It was not yet ready to retire.

Since the Halley Armada and the encounter with Grigg-Skjellerup, there have been more probes sent to explore comets. The NASA Deep Impact mission collided (deliberately) with Comet Tempel 1 (formally 9P/Tempel) in July 2005 and, as it approached, imaged the comet in fine detail. The European Space Agency's Rosetta probe is scheduled at time of writing to land on comet 67P/Churyumov-Gerasimenko in 2014.

In January 2004, NASA's Stardust mission not only encountered 81P/Wild (known as Wild 2, pronounced Vilt-2) but brought back particles collected from the comet's coma, adhering them for the return journey with aerogel. It was an extraordinary achievement, and the analysis of the particles has both widened our understanding of the nature of comets and posed new questions. Wild 2 contained a mixture of materials, some requiring high and others low temperatures to form. This implied that it came together in different locations, not only in the sort of cold environment and locations where ices would condense. There were abundant hydrocarbons in the returned sample.

A Hint of Life

In 2009, it was confirmed that one of these hydrocarbons was the amino acid glycine.

Glycine is used by living organisms as one of the building blocks for proteins. It was thought initially that the glycine might have been a contaminant but has since been provisionally identified in the interstellar medium and, when the contamination theory was ruled out, some interesting conclusions emerged. According to Dr. Carl Pitcher, director of the NASA Astrobiology Institute: "The discovery of glycine in a comet supports the idea that the fundamental building blocks of life are prevalent in space, and strengthens the argument that life in the universe may be common rather than rare."

The connection between comets and life is not new. Halley and Newton both speculated that a comet impact on Earth might have played a role in the planet's early history by transporting water across the wastes of space. There is good reason to think that comets enriched the atmospheres of the giant planets. There is also some evidence suggesting that comets added gases to the noxious brew around Earth's near twin Venus and have contributed to the presence of water in Mars' past.

What about Earth? Did water come here via a comet? And if water, then what about amino acids such as glycine, the building blocks of life? There are hints that collisions between comets and the terrestrial planets can and have occurred. The Rosetta mission will produce a full inventory of organic chemicals in Comet 67P/Churyumov-Gerasimenko, adding new information and no doubt posing new questions as we explore the role of comets as transmitters of life.

Proving the Invisible

Comets are beautiful and interesting in their own right, but when we use predictions of cometary flybys, we are repeating one of the great proofs of scientific history, a moment fundamental to our understanding of the forces of the universe. Comets are evidence of how gravity works.

We have become so accustomed to gravity that we forget how strange an idea it is that something we cannot see is binding the planets into elliptical orbits around the Sun. Kepler had perceived in the early seventeenth century that there had to be some force of attraction such that: "If two stones were placed anywhere in space near to each other, and outside the reach of force (of other bodies), then they would come together…at an intermediate point, each approaching the other in proportion to the other's mass."

Newton's early years in Cambridge were a good time to be watching comets. "So to the Coffeehouse," the English diarist Samuel Pepys (1633–1703) noted for December 15, 1664, "where [there was] great talk of the Comet seen in several places." A few days later, there was "mighty talk of this Comet that is seen at nights; and the King and Queen did sit up last night to see it." Another comet appeared in 1665, and in 1668 the great astronomer Johannes Hevelius (1611–1687) produced his *Cometographia*, classifying comets by the appearance of their tails. In 1680 a new comet was spotted by Gottfried Kirch (1639–1710), and Newton used first his eyes and then a monocle and finally telescopes to keep it in view.

When another comet appeared 2 years later, Newton met to discuss his observations with Edmund Halley (1656–1742). Halley was by this time a Fellow of the Royal Society who had published an important catalog of southern hemisphere stars. Halley and Newton kept in touch.

Then came a revelation. We are told that in 1684: "Dr. Halley came to visit him at Cambridge, after they had been some time together, the doctor asked him what he thought of the curve would be that would be described by the planets supposing the force towards the sun to be reciprocal to the square of their distances from it. Isaac replied immediately that it would be an Ellipsis, and the doctor struck with joy and amazement asked him how he knew it, why said he I have calculated it."

In other words, Newton said he had worked out the mathematics of orbits and the gravity that acted on them.

Throughout this time, and especially between 1680 and 1684, Newton was working hard on comets. He collected information from wherever it was available – Europe, Maryland, Brazil and China. He made use of some of the contacts of that voracious collector of data, Athanasius Kircher (1601–1680). Writing to John Flamsteed, Newton declared his intention "to determine the lines described by the comets of 1664 in 1680 according to the principles observed by the planets."

This work grew into one of the great achievements in the whole history of science, Newton's *Philosophiae Naturalis Principia Mathematica*, usually just called the *Principia*. This book, first published in 1687, was where Newton laid out his laws of motion, giving the ground rules for modern mathematics and science. In Book 2, Newton established that: "During the whole time of their appearance, comets fall within the sphere of activity of the circumsolar force and…therefore describe conic sections…. For that force, propagated to an immense distance, will govern the motion of bodies far beyond the orbit of Saturn."

That force was gravity, but where was the final proof that gravity was in command of all celestial objects? Halley, in his *Synopsis of the Astronomy of Comets* (1705), put together the observational data to suggest that comets were so firmly under the influence of Newtonian gravity that their return could be predicted. Halley stated that the comet of 1682 would return in 1758, an orbital period of 76 years.

It was an act of faith in science – in Newton's theories of gravity that refined Kepler's understanding of elliptical orbits and in Halley's own trawl through the history of observing. It was an act of faith post mortem, too – Halley died in 1742.

By 1758 there were both believers and skeptics concerning the prediction. "True lovers of science desire its return," wrote Alexis Claude Clairault (1713–1765), a supporter of Newtonianism in France, "because it would afford striking confirmation of a system in favor of which nearly all phenomena furnish conclusive evidence." The other side were inspired, he said, by "the hope it will not return, and that the discoveries of Newton and his partisans may prove to be… the result of imagination."

The comet was late. But a trio of French astronomers worked out that it would have to be, taking into account the effects of Jupiter. It was on Christmas Day 1758 that the amateur astronomer Johann Georg Palitzsch (1722–1788) saw the prodigal comet in the sky.

This was how Joseph Jerome de Lalande (1732–1807), who earlier had published corrected versions of Halley's cometary tables, expressed the significance of the observation: "The universe beholds this year the most satisfactory phenomenon ever presented to us by astronomy; an event which, unique until this day, changes our doubts to certainty and our hypotheses to demonstration. [It] places this law amongst the number of the fundamental truths of physics, the reality of which is no more possible to doubt that the existence of the bodies which produce it."

The effect of Newton's discoveries were profound. Astronomy would henceforth not only observe complex orbits but also derive them mathematically using an understanding of the forces at work. The Jesuit priest Roger Boscovich argued in his *Dissertation on Comets* (1746) that the true shape of the universe was not Copernican or Tychonic but in truth Newtonian. When he came to England in 1760, he visited both Newton's tomb and his rooms in Cambridge, a scientific pilgrimage.

The Man of Science

Isaac Newton can appear to be a contradictory figure. Much as some histories of science try to ignore it, he devoted at least as much time to theology and alchemy as to mathematics and astronomy. Alchemy may well have been his deepest interest.

However, this is not a real contradiction. It is forced on us by two modern misconceptions, first that alchemy is mumbo-jumbo and second that knowledge should be organized into distinct and warring faculties, each with their particular heroes.

Whereas for Newton, all the subjects were important. He spent much time studying the Bible, and his researches on prophetic texts alone form a considerable archive. They led him to unconventional religious views, but then Newton was original in all he thought. His alchemical explorations were similarly massive – he wrote 15,000 words of notes on his alchemical experiments between 1679 and 1684 alone, representing many hours spent with alembic and weighing scales. After 1687 he devoted attention to hieroglyphs and occult writings, translating and incorporating them into his thought.

What did Newton believe these studies achieved? For him, theology and alchemy were, with comets and mathematics, expressions of one search after the truth, what he called *prisca sapientia*, original wisdom. "All things come from one root," he wrote in 1669, a creed uniting his studies. He conceived of a first human religion, rational and scientific with places of worship arranged to imitate the shape of the Copernican Solar System. The Sun lay in the center, represented by a fire, and it was attended by priests circling it as might the planets. Newton castigated bad astronomy, as he did bad religion, calling it idolatry. False science and false faith had diverted humanity from true understanding. So his work, whether on the laws of motion of or comets, formed a piece with his alchemy and his theology. His was a mind forever in search of the full truth.

The Art of Comets

Comets, the most spectacular of celestial surprises, have never been neutral in human history. Comets have made kings and have dethroned them. Shown here is the famous image of the English King Harold cowering in fear in early 1066 at the sight of the omen passing over his palace.

Fig. 12.6 The comet scene from the Bayeux tapestry (Credit: public domain)

The comet is, of course, Halley's. This is the earliest image we have of it. In April 1066, it was around 16 million km from Earth and shining at a magnitude of −2. This scene comes from the Bayeux Tapestry, a visual story of the Norman conquest of England created in the 1070 s. The 'hairy star' was a sign from the heavens that Harold's claim to the throne of England was false, so the tapestry makers want us to understand – the comet clearly supported the Norman side, and the Normans won.

A later appearance of Halley's Comet was also captured in art, this time in one of the great paintings of Giotto. Giotto di Bondone (c.1267–1337) was a Florentine artist regarded as a master of the Italian Renaissance. He included a comet in his *Adoration of the Magi*.

Fig. 12.7 The star of Bethlehem in Giotto di Bondone's fresco 'Adoration of the Magi' in de Scrovegni Chapel in Padua (http://www.springerimages.com/Images/Physics/1-10.1007_ 978-1-4419-7811-0_1-23)

Seen here in close-up, it is a remarkably naturalistic rendering of the celestial messenger. Instead of resorting to a conventional pointed star or using an astrologi- cal image, Giotto chose to paint the comet as it appeared, including bright center corresponding to the comet's coma and then a long tail.

Did Giotto draw this image from his own observations? It feels as though he did. It was a revolutionary image to include in a piece of sacred art, a response to a contemporary event that would have had all those around him staring up into the sky and wondering what the apparition meant. Ours today are different questions, for sure, but we still look and we still ask.

Chapter 13

Wonderful Demons

Observation: Algol, M34, Delta Cephei, Mu Ceti and Mira
Significance: Stars vary in brightness, standard candles
Science: Variable stars

In 1844, the German astronomer Friedrich Wilhelm August Argelander (1799–1875) issued this plea: "Could we be aided in this matter by the cooperation of a goodly number of amateurs, we would perhaps in a few years be able to discover laws in these apparent irregularities."

He went on: "Therefore do I lay these hitherto sorely neglected variables most pressingly on the heart of all lovers of the starry havens. May you become so grateful for the pleasure which has so often rewarded your looking upward, which has constantly been offered you anew, that you will contribute your little mite towards the more exact knowledge of these stars!"

Argelander combined his Romantic era writing style with practical astronomy. He was involved during the 1850s in creating the Bonner Durchmusterung catalog of stars that referenced more than 300,000 stars of low apparent magnitude. No wonder his appeal was heard.

We are going to make some of the key observations that have shaped variable star astronomy, interspersing these with glimpses of pivotal moments when variable stars have determined our understanding of the skies. This cannot be a complete survey of every type of variable star, but in a single chapter we will gain an impression of the challenge and opportunity of observing variables, as well as why they matter.

What is a variable star? They are stars that show variations in brightness over short or long periods of time. Some fluctuate regularly and others unpredictably. John R. Perry expresses the interest of seeing these members of the stellar family in his *Understanding Variable Stars* (2007) like this: "Variable stars are 'speaking'

M. Marett-Crosby, *Twenty-Five Astronomical Observations That Changed the World: And How To Make Them Yourself*, The Patrick Moore Practical Astronomy Series, DOI 10.1007/978-1-4614-6800-4_13, © Springer Science+Business Media New York 2013

to us. Variable star astronomers seek to learn their language, and understand what they are saying."

The labeling of variable stars began with Argelander, who assigned them capital letters from R to Z in front of Latin constellation labels. He did this in order – R Orionis (R Ori) is the first of his variable stars in Orion. When Argelander had exhausted the 9 letters, he doubled them, starting with RR to RZ and then SS to SZ all the way around the alphabet back to QZ. QZ Orionis is the 334th variable star in Orion. Argelander started with R because R stood for the German *rot*, meaning red.

If a constellation requires more labels, astronomers now use a single uppercase V plus numbers greater than 334.

Distinguishing Variables

Variable stars speak a range of different dialects, and it helps to make distinctions. Some of them vary because of their composition – we label these intrinsic variables – and others because of something that occludes them, perhaps by blocking periodically the line of sight from Earth or because of surface features on the star. These are the extrinsic variables.

Each of these two groups is then further divided. The distinctions often rest on details beyond our power to observe. Old systems become obsolete. New paradigms emerge. Each class of variable is usually named after a model star, either the first to be identified or the one that displays most prominently the defining feature.

Meet the Demon

The study of extrinsic variables begins with Algol, the beta star of Perseus. It gives its name to a conundrum, the Algol paradox, and reveals one of the strangest interactions between stars in the sky.

Algol's changes in brightness are visible to the naked eye. Early astronomers were drawn to this and perhaps frightened by it. A star of many names, our modern Algol is derived from the Arabic for both Demon's Head and Mischief Maker, and it lies at the root of the English word *ghoul*. It rules a part of the sky once known as Caput Larvae, The Spectre's Head. In Hebrew, it was connected with Satan. Its Chinese title translates as The Piled-Up Corpses.

The demon star rests in the grasp of Perseus, the hero constellation that reigns over much of the northern night sky, but it is also visible in southern latitudes above 35° in the summer. We start by finding the brightest star in the constellation, Alpha Persei, or Mirfak. Draw a line between Pollux and Capella (Alpha Aurigae), and it is the next bright star.

Mirfak is a supergiant star in the F spectral group with a troubled atmosphere. Through binoculars or a telescope the star becomes the bright heart of a cluster of young and massive stars called Melotte 20, or the Alpha Persei Cluster, an evolved group of stars some 600 light years from us and sharing a common origin. There are some lovely siblings roundabout, including the soft orange star Sigma Persei very close to the main star.

Fig. 13.1 Melotte 20 with Alpha Persei (Mirfak). Canon EOS 5D, ISO 1000, Canon EF 200/2.8 @4.0, four 30 s and two 15 s exposures (Credit: Peter Wienerroither, http://homepage. univie.ac.at/peter.wienerroither/)

Move a few degrees southwest to find Algol. It's not the first star that you come to but the second. The variations in Algol can be seen with the naked eye, but they're easiest in comparison with a steadier, unblinking star. If you are using binoculars or have a wide field of view, make use of the bright star that makes a clear triangle with Mirfak and Algol. This is unvarying Gamma Andromedae, Almaak, which at magnitude 2.15 provides an indicator for assessing its neighbor.

Originally named after the caracel, a hunting cat, the Arabic name for Almaak was misread during the European Renaissance as meaning *boot*. Our modern Almaak (also Almach) is a conflation of the cat and the boot.

Algol varies in brightness between a minimum magnitude of 3.4 and a maximum of 2.1 over a rapid cycle of just 2 days and 21 h. This is best observed frequently over a short period. Taking Almaak as a fixed point, note down for each observation whether you see Algol shining more or less brightly than its neighbor, and then refine your judgment by consulting one of the online sources that give precise readings for its magnitude. Such lists almost always use Universal Time and the Julian calendar, so you will need to convert your local time and date. For best results, get Algol into view about 2 h before the predicted minimum and then return periodically. The star is at its minimum for 20 min and then gradually starts to brighten again. The demon star is winking.

Before we explore the science of Algol, there are some must-see observations in this part of Perseus. One is the star just a degree or two below Algol, fire-colored Rho Persei. It fits into the same binocular view with Algol, and they make a fine contrasting pair.

Search out M34, which is the third corner of a triangle stretching away towards Andromeda with Algol and Kappa Persei. A small telescope will reveal 30 or so stars, and the larger the instrument, the more you'll find. It is a magnitude 5.2 loose cluster in the shape of a cross, with lots of double stars within the group. Keep the heart of the cluster in the center of the image and increase magnification to see the group grow in number. It might be as much as 225 million years old – the stars of M34 turn mighty Mirfak and Melotte 20 into the boastful new kids on the block.

A hint for northerners viewing Perseus. It is, literally, a pain in the neck. Being so high overhead in the winter, the constellation is too high for using the normal props that sustain binoculars, and with hands alone the wobble, especially in the cold, can ruin the sight. Odd as it sounds, a Sun lounger makes for an ideal platform for viewing Perseus – lie back and enjoy. We will return to Perseus in a later chapter to view the celebrated Double Cluster.

What is happening to make Algol wink? It is a B-type star on the main sequence of the HR diagram, fusing hydrogen like our Sun. It is 3½ times more massive and, at a distance of 93 light years, has an average visual luminosity about 100 times greater. There is nothing intrinsically peculiar, then, about Algol itself.

The star was observed for many centuries before an 18-year-old deaf mute from Yorkshire, John Goodricke (1764–1786), offered the explanation. Writing to Antony Shepherd, Plumian Professor at Cambridge, Goodricke proposed: "I should imagine it could hardly be accounted for otherwise than either by the interposition of a large body revolving around Algol, or some kind of motion of its own, by which part of its body, covered with spots or such like matter, is periodically turned towards the Earth."

Goodricke offered two explanations here. The second was the one generally accepted at the time, while the first provoked some interest but was soon forgotten. Not much notice was given to Goodricke's idea of something else in orbit around Algol that occluded it until it was revived first by T. S. Aldis in 1870 and then by Edward Pickering in 1881.

Goodricke's first instinct was right. Algol is the prototype eclipsing variable, and studies using photometry and spectroscopy have now demonstrated that the main star, Algol A, is circled by a companion, Algol B. Algol A shines with a surface temperature of 12,000 K, while Algol B has a surface temperature of 4,900 K.

Each time the dimmer star passes in front of the brighter, the total output of this binary pair drops by an observable amount. There is then a smaller secondary dip when Algol A passes in front of Algol B, occluding its lesser contribution to the luminosity of the system. There is also an Algol C, but it plays no part in the drama.

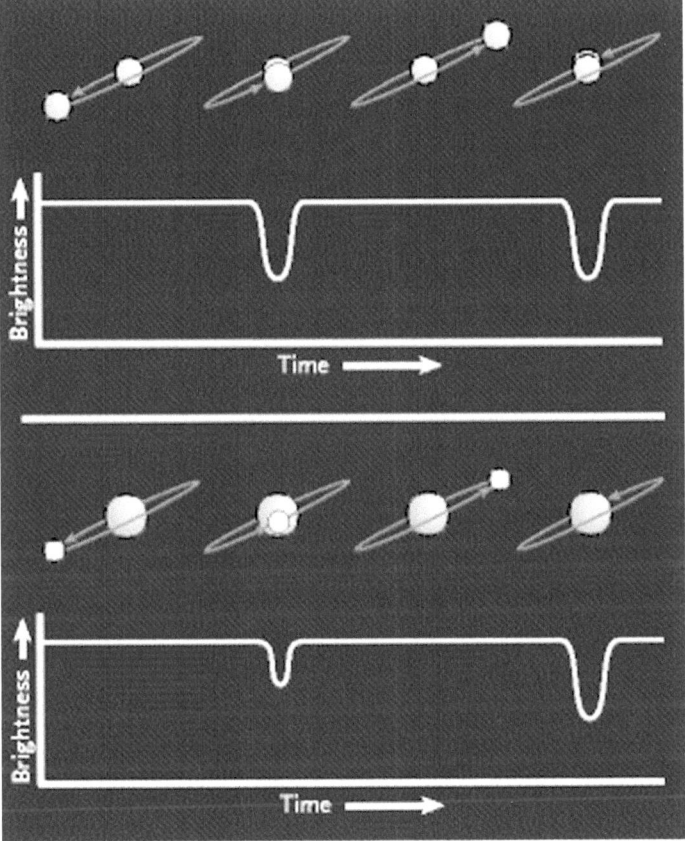

Fig. 13.2 Two eclipsing variables: *top*, a binary star composed of similar stars (for instance, Beta Lyrae): *bottom*, binary star composed of an orange giant and a white hot star (as in Algol) (http://www.springerimages.com/Images/Physics/1-10.1007_978-0-387-85355-0_3-7)

This is not the sum of Algol's mysteries. The theory of stellar evolution states that the rate of a star's change depends on its mass, huge stars moving more quickly through the HR diagram towards giant life and cataclysmic death. But binary stars such as Algol do not work like this, the so-called Algol paradox. Something is keeping it alive. The paradox is resolved by the existence of a flow of material between the two stars, their physical bond so intimate that the more massive Algol A feeds off Algol B. Known as Roche Lobe Overflow, models suggest that mass from Algol B collects in a disk around Algol A and then spirals into the heart of the larger star. The secondary star is the older one, but smaller because of its mass transfer.

Of Kings and Queens

There are two targets for this next variable star observation, both of them residing in the palaces of royalty. We will be looking for classic variable stars in the constellations Cepheus the King and Cassiopeia the Queen. The whereabouts of Cepheus, once king of the Ethiopians, is best found by locating his queen, who has at the center of her constellation a w-shaped asterism of five stars. This asterism makes it straightforward to delve into the constellation Cepheus the King. The arm of the w further from Polaris points towards the constellation.

Track across a comparatively empty slice of night along the line these stars make, and you will reach a triangle of stars in an isosceles pattern. The star at the peak of the two long sides of the triangle, which is the first star you reach coming from Cassiopeia, is our next observing target, the star Delta Cephei. It functions with the reliability of the most perfect clock, its magnitude moving in between 3.5 and 4.3 in precisely 5 days 8 h 47 min and 32 s. Use the fixed brightness of Zeta Cephei beside it, shining at magnitude 3.6, and take some notes on a star that has given its name to the whole family of Cepheid variables.

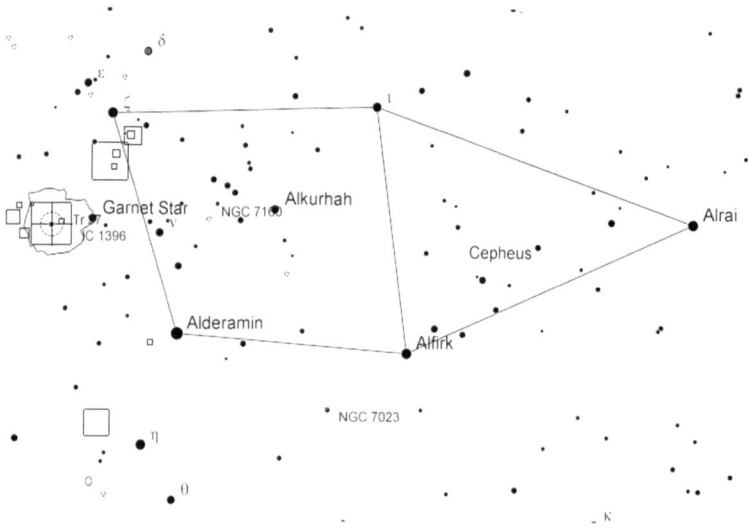

Fig. 13.3 The Constellation Cepheus (Cartes du Ciel)

What is happening in Delta Cephei? Cepheids are all huge and dying stars. Their cores have exhausted their supply of hydrogen and are moving into a more advanced phase of nuclear fuel consumption and burning helium. They are subject during this phase to vast radial pulses, convulsing the star as they expand and contract with the regularity of beating hearts. Their regularity and abundance have given them a special place in astronomy.

Compare Delta Cephei with its near neighbor Epsilon Cephei, another variable star but of a very different type. It belongs to the Delta Scuti family, a shrunken version of the Cepheid type with multiple patterns of variation, all very small.

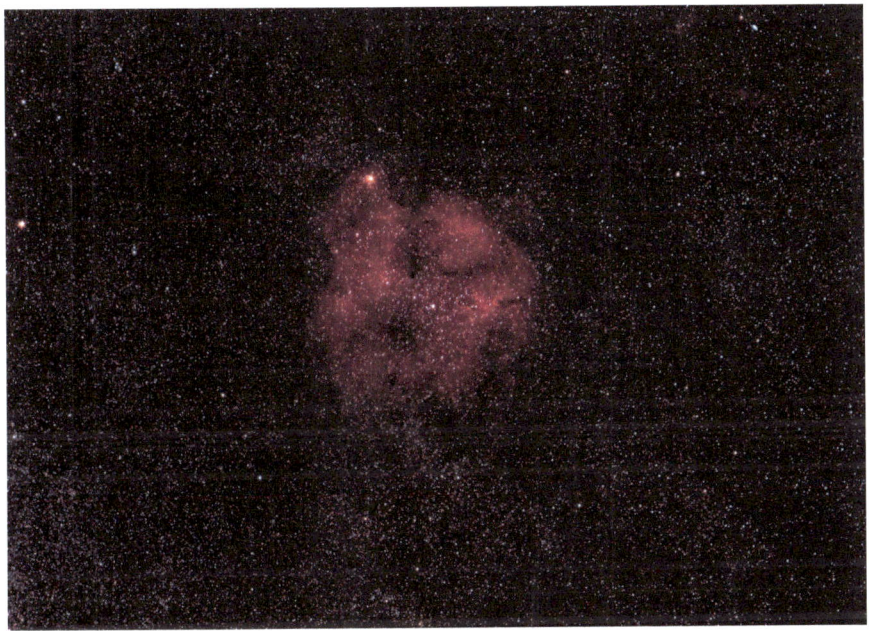

Fig. 13.4 Mu Cephei and IC1396 shot with an EOS 350D mod, 800ASA, 135 mm lens f/4, HaGB, 9×8 min and 15×2 min, piggyback on ETX-90RA from Lower Austria in August2008 (Credit: Erwin Matys and Karoline Mrazek)

While in the presence of the king, it is worth finding Mu Cephei, the Garnet Star. "A very fine deep garnet color," was how Sir William Hershel described it in 1783. From Delta Cephei, move just a few degrees away from the Cassiopeia asterism and you will see it with the naked eye and certainly with binoculars. Some call it red and others orange or even purple.

Mu Cephei is probably the largest star visible to the naked eye. Some 1,600 times larger than the Sun, in our Solar System it would extend almost as far as Saturn. It is also another prototype variable star, identified as such by John Russell Hind in 1848, and it probably has just a few million years to go before it explodes.

Alpha Cephei's modest magnitude of +2.5 makes it pale for a Bayer alpha star, but it is still visible to the naked eye. At 49 light years away from us, it is a variable in the Delta Scuti class with an interesting future and a puzzling present. In five and a half millennia or so will be pretty close to the celestial north pole. Meanwhile, Alpha Cephei is rotating at very high speeds, as much as 246 km per second, with the result that the star has an oblate shape, fatter at its equator than on the axis of its poles.

A study using the CHARA array at the Mount Wilson Observatory has created an image of the star showing this elongation in the east–west direction. Also detected were bright and dark regions on the surface of the star, a phenomenon known as gravity darkening. The deep processes at work in stars like this are not fully understood, but close observations are starting to unlock another part of the science of stellar evolution.

In the muddled world of star labeling, Alpha Cephei may win the prize. The original Arabic was translated as "the King's right arm." By the time Bayer published his *Uranographia* in 1603, its name was Aderaimin or Alderamin. It variously became Adderoiaminon, Assemani, Alderal jemin, Al Derab, Al Deraf, Al-redaf, and Alredat. Other Arabic sources label it Al Firk, meaning a flock, perhaps echoing a label of the great fifteenth century astronomer-king of Samarkand Ulug Beg, who called the bright stars in this area the Stars of the Flock or possibly the Herd of Antelopes. Alderamin is the nearly settled name used today.

Standard Candles

Stars such as Delta Cephei transformed the shape of space, enabling us to measure distance as never before. Stars in this family of variables made possible Edwin Hubble's discovery of the extent of the universe.

We have distinguished before between apparent brightness and intrinsic brightness, the latter a measurement of the power a star is radiating into space. There is a relationship between them, and the difference between them is going to have a lot to do with distance. If we can establish a value for luminosity, then distance will pop out of the equation.

Enter Cepheid variables and the work of Henrietta Swan Leavitt (1868–1921) at the Harvard College Observatory. She was employed there by Edward Pickering to count images on photographic plates, paid at a rate of $10.50 a week. Leavitt did her counting and a lot else besides. She looked at 1,777 variable stars in the two Magellanic Clouds of the Southern Hemisphere, where she worked out that in the Delta Cepheids, there was a predictable relationship of luminosity to period. In her words: "A straight line can readily be drawn among each of the two series of points corresponding to maxima and minima, thus showing that there is a simple relation between the brightness of the variables and their periods. The logarithm of the period increases by about 0.48 for each increase of one magnitude in brightness."

This was the key to unlocking the distances between the stars. Leavitt's work meant that, for any galaxy containing a Cepheid, it was possible to deduce magnitude from the observable period and then, by rearranging the equations, emerged with a value for the elusive, vital measurement of distance.

The equation for absolute magnitude is $M = m - 5\log d + 5$, where absolute magnitude is M, apparent magnitude is m and d is distance.

On New Year's Day 1925, Edwin Hubble informed 4,000 scientists at the American Association for the Advancement of Science that this was what he had done. His paper, *Cepheids in Spiral Nebulae*, applied Leavitt's discovery to Cepheids in the Andromeda and Triangulum galaxies and offered the conclusion that these stars were situated far beyond the borders of the Milky Way. Like light-houses, the flashing Cepheids had shown the way.

Hubble was not there to share the moment. He had returned to California, con-cerned that someone would raise an objection he had not thought to consider. He had worked with certain assumptions, notably that there was no distorting medium within the spirals to confuse luminosity measurement and that Cepheids always behaved in the same way. At 35 years old, he was young to change the world. No wonder that he felt a little anxious.

A study undertaken with Hubble Space Telescope data gathered between 1995 and 1997 puts some numbers onto his revolution. The evidence identifies 32 Cepheids in the Triangulum (M33) and uses Henrietta Leavitt's straight line to place the galaxy around 2.74 million light years away. This is how big the universe is or, to invert the mathematics, how small our home turns out to be.

Fig. 13.5 M33, The triangulum galaxy field of view approximately 2×2°. Software used: Fits Liberator, Photoshop, RegiStar (Credit: John Corban & the ESA/ESO/NASA Photoshop FITS Liberator)

Cepheid variables have become known as 'standard candles,' reliable milestones for measuring distances to the stars. They still perform this important function in astronomy. It is worth noting, though, that even Cepheids sometimes turn out to be unreliable. One of Hubble's own stars, V33 in the Triangulum, is no longer pulsing as a proper Cepheid should. The same puzzling result has been found for Polaris. "I am as constant as the Northern Star," Julius Caesar proclaims in Shakespeare, "of whose true fixed and resting quality, there is no fellow in the firmament." It turns out that this may not mean quite what the Bard intended.

Whale Watching

Cetus the Whale is a broad, straggling constellation with a lot of not very bright stars. It covers a vast area of the night sky during the northern winter, when it's best observed late in the night. A good starting point is the Pleiades star cluster, which is easy to find at the opening of a crisp winter night of whale watching. The constellation Cetus lies below the Pleiades and to the east of Orion and the bright star of Taurus, Aldebaran.

With this general sense of the star field in mind, it's helpful to return to Aldebaran and identify the arrowhead shape of the stars above and around it. These stars, though not Aldebaran itself, belong to the Hyades star cluster, about which there will be more in a later chapter. The tip of the arrow is Gamma Tauri, and the Hyades asterism is pointing straight towards Alpha Ceti. Follow the line by way of one star only, and you're there.

The alpha star of the Whale is the somewhat inconspicuous Menkar shining at magnitude +2.5. In binoculars, the star 93 Ceti looks like a companion to Menkar, although in fact they are hundreds of light years apart. Mira lies on an extension of a line between Menkar and Aldebaran that passes through a fairly bright star, Delta Ceti, on the way. Trace the route in binoculars and then repeat the process using a finderscope to tie the telescope onto the star. Just a few degrees below Delta Ceti you will find Mira, Omicron Ceti.

The Whale has a confused anatomy. Menkar means "nostrils," and is applied to Alpha Ceti as a result of a mistake in a medieval star chart. The nostrils of the Whale originally lay upon Lambda Ceti and not Alpha.

The Wonderful Star

What is Mira? The star was first noted by David Fabricius (1564–1617) while he was hunting for the planet Mercury. Returning to the star after a few months, it seemed to have disappeared. Just a few years later, Bayer recorded it for his *Uranometria* of 1603, but it was Johannes Holwarda (1618–1651) who established that Omicron Ceti wobbled irregularly between a bright magnitude 2 and a very faint 10.1 over period of 11 months.

Mira-type stars, and there are many that share its characteristics, have periods of between a hundred and a thousand days, and exhibit pronounced variations in magnitude of more than 2.5. Mira itself is a large, cool red giant, as are all stars in this class, making them relatively easy to spot, as despite their low temperatures their huge diameters mean that they shine many thousands of times more brightly than our Sun. Unlike Cepheids they cannot be plotted along a tidy straight line. In some stars of this class, maximum brightness can be very different from one peak to the next. A star such as R Andromedae, a challenging target not far from the Andromeda Galaxy, exhibits unpredictable variations. It is altogether an extraordinary star, the first to reveal the presence of the rare and unstable element technetium in its spectrum.

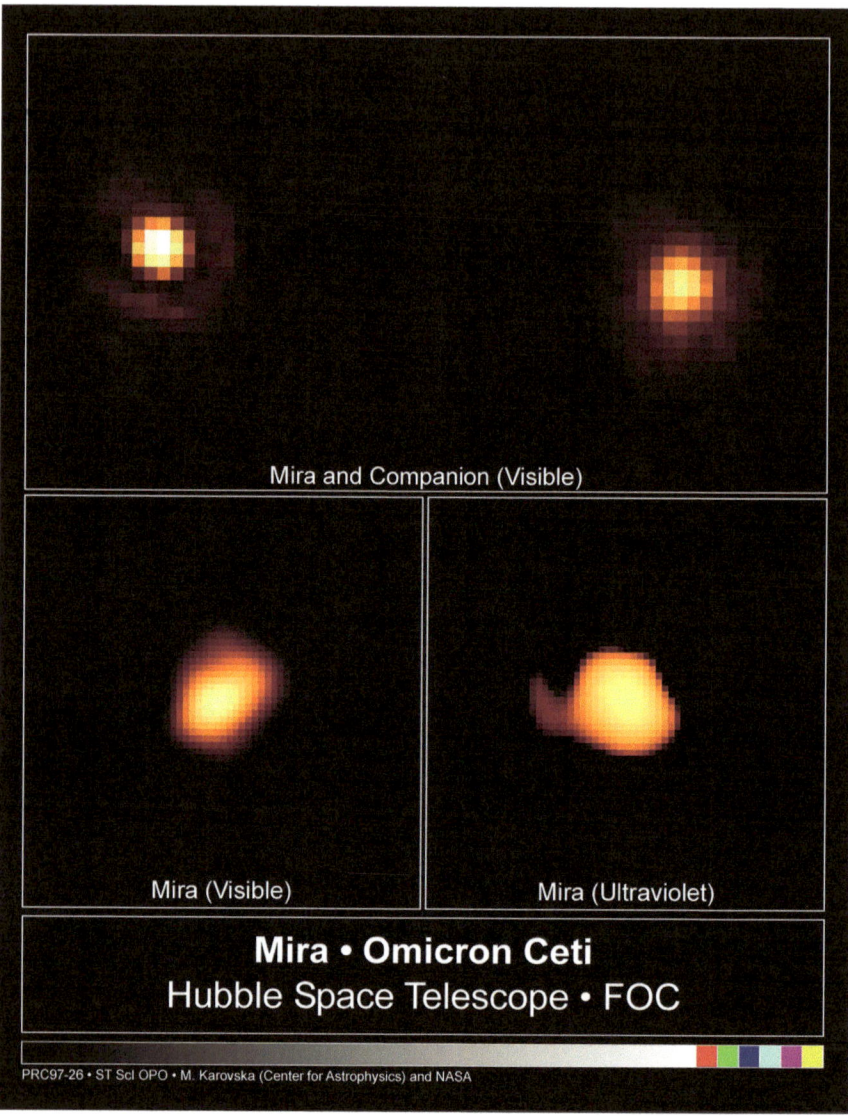

Mira and Companion (Visible)

Mira (Visible)

Mira (Ultraviolet)

Mira • Omicron Ceti
Hubble Space Telescope • FOC

PRC97-26 • ST ScI OPO • M. Karovska (Center for Astrophysics) and NASA

Fig. 13.6 (*Top*) NASA Hubble Space Telescope image of the cool red giant star Mira A (*right*), officially called Omicron Ceti in the constellation Cetus, and its nearby hot companion (*left*) taken on December 11, 1995 in visible light using the European Space Agency's Faint Object Camera (FOC). (*Lower left*) Hubble visible-light image of the disk of Mira. (*Lower Right*) Hubble UV light image of the disk of Mira (Credit: Margarita Karovska [Harvard-Smithsonian Center for Astrophysics] and NASA)

What is going on inside Mira variables? The image of Mira shown taken by the Hubble Space Telescope in 1995 reveals that the star is far from circular. Something dramatic is causing this, presumably connected to its pulsations. Irregular shapes are common in stars of this type. So are high rates of mass loss and the creation of rings of circumstellar material. Mira lies in the Asymptotic Giant Branch (AGB) of the HR diagram above and to the right of the main sequence.

The Hubble images also reveal that Mira has a companion star, separated by about 70 times the distance between Earth and the Sun. The bottom right of the four pictures shows an arm reaching from Mira towards Omicron Ceti B. This might be material being exchanged between Mira and the smaller star or an atmospheric effect. Observations made in 2010 of Mira reveal arc-like structures in its ejected material caused by interactions with the other star and the interstellar medium.

Mira-type stars reward careful observation, and records of their activities by amateur astronomers make a genuine contribution to science. This is often done through associations such as the American Association of Variable Star Observers or the Variable Stars section of the British Astronomical Association. Both of these run websites with predictions, light curve tools and downloadable observing forms. This returns us to where we started and Argelander's appeal of 1844 – we can all add our little mite.

Night-Gaunts

Before we become too comfortable that we comprehend all that we see in space, one might read the science fiction of Howard Phillips Lovecraft (1890–1937).

Lovecraft is a great exponent of weird fiction. He expresses views on race that are deeply unattractive, and he rarely includes a single female character in his fiction, but these were part of the palette of his age and should not blind us to his work. He wrote out of his dreams, he said, "nightmares of the most hideous description, peopled with things I called night-gaunts." Lovecraft's short stories are studies in the unfathomable, tales of collapse in which human beings are tiny creatures in a cruel universe.

Thus in "The Colour Out of Space" (1927), a surveyor is sent west of Lovecraft's fictional Massachusetts city of Arkham and finds the site of a projected reservoir to be first "bad for the imagination" and "no region to sleep in." The cause of this is a story of a meteor fall, "strange days" remembered by Ammi Pierce that began with a white noontide cloud, a pillar of smoke and a rock falling to Earth on the trim farmland of Nahum Gardner. Professors came to examine the rock, which had a color that was real but could not be described. "The professors felt sure they had indeed seen with waking eyes that cryptic vestige of the fathomless gulfs outside; that lone, weird message from other universes."

The scientists failed to understand it. The countryside grew sick. "A stealthy bitterness and sickishness" crept over plants and trees and people. Bodies changed shape while "the dark fears of rustics were held up to polite ridicule." Everything lay under that diseased color, the one "without a place among the known tints of earth." Finally, a cloud creeps over the Moon, and "a thousand tiny points of faint and unhallowed radiance" alight onto the treetops, while "the shapeless stream of unplaceable color… seemed to flow directly into the sky." It melts back into the Milky Way. The narrator says of what he has heard: "Do not ask me for my opinion. I do not know – that is all."

At stake in this story is a stable structure to reality. Colors, science tells us, live along a spectrum. But from outside comes a color that does not fit category or language, "insolent in [its] chromatic perversion." The people fail to salvage their certainties. Even the Biblical is undone – the meteorite is a pillar of cloud and light, but it does not guide. This anti-Pentecost reduces humanity to an incomprehending silence.

Lovecraft's cosmic horror is a reaction to the certainty that we will comprehend everything we see. His vision of human beings is grim, at times appalling. He says of one of his dreadful creations that "They're like us; we are these creatures. The horrible things that they're doing are things that we do." Meanwhile his scientists are truthful. They say that they do not always know, which is the truth.

Chapter 14

The King of the Planets

Observation: Jupiter
Significance: Planetary accretion, the speed of light
Science: Nature of gas giants, L, T and Y dwarf stars

In an earlier chapter, we turned our telescopes towards the satellites of Jupiter. In this we were following Galileo's example, making an observation that changed the place of humanity in the Solar System. So why go back to Jupiter? In part, because the planet is an enduringly enjoyable observation, with lots of detail to observe, sketch and image. It is also the nearer of the two gas giants, and its structure and processes, very different from our own, demonstrate how all the planets came together by accretion. It also stretches our definition of a planet, introducing us to new kinds of stars. Above all, it was by careful observation of Jupiter that astronomers first identified evidence for light having a finite speed.

Observing Jupiter

After the Moon, the astronomical object that reveals the most detail is Jupiter, the giant planet of our Solar System. Containing some 70 % of the mass of all the planets combined, its travels through the night sky are spectacular.

Jupiter is not always visible from Earth. As it travels on its elliptical orbit, an average of 5.2 AU from the Sun, it can be concealed in the glare of daylight. But astronomy magazines provide clear guidance as to when the planets are visible. So do many websites and computer programs.

Jupiter takes its name from the king of the Roman gods and is formed from the name *Jovis* and the word *pater*, meaning "father." The analog to Jupiter in the

M. Marett-Crosby, *Twenty-Five Astronomical Observations That Changed the World: And How To Make Them Yourself*, The Patrick Moore Practical Astronomy Series, DOI 10.1007/978-1-4614-6800-4_14, © Springer Science+Business Media New York 2013

Greek system is Zeus, and the non-Galilean satellites of Jupiter are mostly named after gods with some relationship to Greek Zeus. The adjective that refers to the things of Jupiter is Jovian.

The planet is always worth finding, if only with the naked eye. At opposition, shining at magnitudes of −2 to −2.9, Jupiter is magnificent. A small telescope shows dark stripes, the North and Southern Equatorial Belts in the Jovian cloudtops. Through larger telescopes, more belts and the gaps between them, known as zones, start to appear. The image shown, taken through an amateur telescope, shows a wealth of bands and zoning on the planet, as well as capturing bright Io alongside it.

(For more information on recording your observations of Jupiter, see the American Lunar and Planetary site http://www.alpo-astronomy.org/jupiterblog and the British Astronomy Association's http://www.britastro.org/jupiter/.)

Fig. 14.1 Jupiter and Io (Credit: Roger Homan, http://www.delscope.demon.co.uk/)

Beginning from the planet's South Pole, which is at the top in telescopes that do not have an erecting prism, it is possible with decent aperture to observe the South Polar Region, the South South Temperate Belt and Zone, then the South Temperate Belt and Zone and the South Tropical Zone before reaching the most visible Equatorial Belt. The same pattern is then repeated for the Northern Hemisphere.

But contrast on Jupiter is a matter of subtlety. High magnifications make the planet watery and opaque. To get closer to the Jovian features, it is far better to make use of color filters. These don't create colored, Hubble-like images, but they enhance contrast. A red filter will draw the eye to white patches in the South Temperate Belts and Zone and bring out pretty features along the edges of the Equatorial Belt. A blue filter adds detail in the brighter zones and brings into prominence one of the prizes of Jovian observing, the Great Red Spot.

Most observers develop their own pattern of filter use, and it's important to keep trying new ones. Jupiter is an ever-changing target, both for the patterns made by its satellites but also because of the planet's own rotation. Particular belts and zones can fade or even disappear, so it is valuable to keep a record of Jupiter observations. There are observing forms available online, and a system of notation for key features.

Another interesting observing target for Jupiter concerns the shape of the polar regions. It's possible in the right conditions to identify a distinct flattening of the shape of the sphere. This is not an atmospheric distortion but a sign of Jupiter's rotation and composition. Different latitudes of the planet rotate at different speeds, and as Jupiter is mostly gas, the effect is a noticeable squashing. The planet's diameter through its poles is over 9,000 km less than its diameter as measured through the equator.

Observing the Great Red Spot is one of the treats of amateur astronomy. Greater in itself than the diameter of any terrestrial planet, it is a vast and ferocious storm rotating anticlockwise in the Southern Equatorial Belt, completing a full rotation every 6 days. It seems to change in color and brightness – another good reason to keep records – as cloud formations develop around it. Other storm features are also sometimes visible in the planet's atmosphere, though none has the longevity or sheer scale of the Great Red Spot. It has been observed for certain since 1821 and may be the same mark that Robert Hooke spotted in 1664. The Great Spot is a devouring monster – a new red spot that appeared on Jupiter in May 2008 was consumed by the greater storm later in the year, its demise captured by the Hubble Space Telescope.

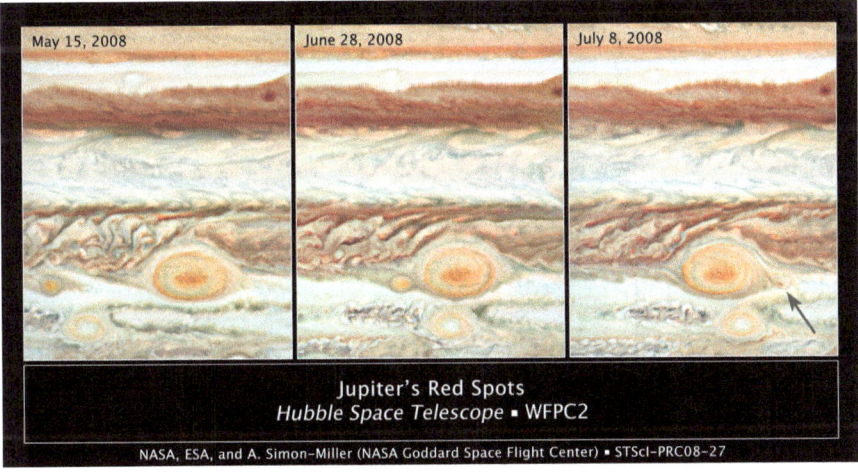

Fig. 14.2 Death of a storm in the cloud-tops of Jupiter

What Is a Gas Giant?

Jupiter, together with Saturn, is classed as a gas giant. This label distinguishes them from the giant planets Uranus and Neptune, which are the subject of a separate chapter. In what follows, much of the science is shared by Jupiter and Saturn, but we will note some important differences when we look at Saturn in a chapter to come.

In observing Jupiter, we are looking at the upper surface of its atmosphere. Nothing we see is solid. There is no way at present of exploring gas giant interiors directly, and so science proceeds by constructing models developed from measurements of density, magnetic fields and the heat emitted. This last point is important – both Jupiter and Saturn emit more heat than their distance from the Sun should permit, indicating powerful activity within.

The densities of the gas giants are very low. But the mass of material forming the planets is very large. Jupiter is 317 times and Saturn 95 times more massive than Earth. Density measurements support a model for both planets that has at its center a core some 50 million times denser than the outer rims of the planets. This core, perhaps 5–10 times the mass of Earth, is probably composed of materials like iron, silicates and icy matter, and it might be differentiated, with the iron at the center.

Was this core the 'seed' around which the gas giants formed? This is one model for the growth of Jupiter and Saturn by accretion, with a kernel of rock and ice that attracted gas from the Solar System. This is sometimes called the runaway capture or 'bottom-up' model. But it's not the only one. The planets might have started the other way round, 'top-down' as it were, beginning as dense clouds of gas and dust, with the rocky material channeled in through the gas to form a core.

These theories have received renewed attention because we now know that gas giants are no longer confined to our Solar System alone. Most of the exoplanets so far discovered are Jupiter-like giants close to their parent stars. Other models have been proposed that might explain how these exoplanets have ended up so close to their stars, with one suggestion that giant protoplanets form from gravitational instability within the dense gas clouds left over from star formation. Alternatively, it has been suggested that gas giant interiors do not work the way we think and that Jupiter's interior may not be sharply differentiated between core and gaseous layers. This might allow such planets to be much richer in heavy metals, a conclusion that would explain the densities measured in some of Jupiter's cousins in distant systems. There is a potentially powerful cross-fertility of astronomical disciplines here, with exoplanetary observations forcing us to reassess how we understand our 'local' giants.

All the models depend on accretion. Planets do not spring into being – they grow. And what Jupiter and other gas giants have attracted above all are the two most common elements in the universe, hydrogen and helium. At the high pressures existing towards the center of mass, the hydrogen nuclei are probably so squashed that they share electrons, forming the metallic hydrogen that is likely to be the source of the planet's magnetic field. The planet's atmosphere is predominantly hydrogen and helium, with ice clouds made of water, methane and other gases. It is these clouds that we observe as they rotate sometimes at speeds in excess of hurricane force on Earth. These form distinct belts and zones, the darker material sinking and the lighter rising.

There is a good deal of change taking place in these atmospheres, caused mostly by internal processes but also by impacts. One such was observed by amateur astronomer Anthony Wesley on July 19, 2009. Using a 14.5-in. Newtonian, he spotted and then tracked the impact 'bruise.' Other observers confirmed that it occurred between 7:40 and 14:02 UT. Professional telescopes were then able to analyze the elongated streak to determine the angle at which the impactor struck the atmosphere. The mark is visible as a small white circle in the upper hemisphere of this image.

Fig. 14.3 GEM mounted Newtonian using a 14.5″ Royce conical mirror on Losmandy Titan mount; Optics 14.5″ f/5 Royce conical primary, 1/30 wave Antares Optics secondary, Televue 5× powermate working at 7.7×; Camera: Point Grey Research Dragonfly2 mono camera, ICX424al (Credit: Anthony Wesley, www.acquerra.com/au/astro)

This observation confirms how major impacts have modified and continue to modify the shape of our Solar System.

Probing Jupiter

On July 3,1782, a rather unusual astronomical event took place – regal, foggy and a great success. The astronomer responsible was Caroline Herschel (1750–1848), sister to William, who the previous day had demonstrated their telescope to the British royal family. It had gone so well that she had received a return invitation, their various majesties and highnesses hoping to observe Jupiter and Saturn.

This was not the usual astronomical party. Caroline records that the royals were anxious "to see without going out on the grass." A telescope was erected in the queen's apartments. But then the weather intervened. "The evening appeared to be totally unpromising." Many of us know just how that feels.

Caroline Hershel had a solution: "I proposed an artificial object, since we could not have the real thing. I had prepared this little piece beforehand, as I had guessed by the appearance of the weather in the afternoon that we should have no stars to look at."

The "little piece" was a painted picture of a planet erected onto a board. This Caroline Hershel attached to the end of the palace garden wall.

The result? A great success, their highnesses "much pleased" with the event. Caroline was surely relieved, but she was also disappointed: "Nothing will give me greater happiness than to be able to show them some of those beautiful objects with which the heavens are so gloriously ornamented."

The weather still frustrates us. But human science can rise above the clouds, and in the case of Jupiter the development of a science of gas giants has relied crucially on space probes sent to the Jovian system. Of these, the most important has been Galileo, launched via the space shuttle after a series of delays in October 1989. It was judged to be "the most capable exploring machine ever developed."

The Galileo probe's achievements were many. Approaching Jupiter, it suffered a tape recorder malfunction but was nevertheless able to achieve its arrival day tasks: flybys of Europa and Io and then orbital insertion around the giant planet. It then dispatched the atmospheric probe it had been carrying, putting to the test theories of Solar System formation by measurements of the abundances of hydrogen and helium.

The first results were surprising. It suggested that helium abundance was 14 %, much less than the expected value. However as more data arrived, the helium abundance rose to 24 %, completely in agreement with the theories. The probe made detailed studies of the Great Red Spot. It was able to view individual storms, lighting, aurora and anvil-headed clouds.

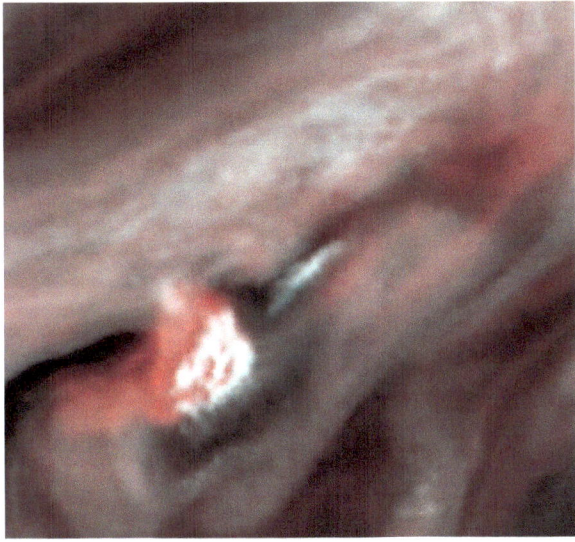

Fig. 14.4 False-color picture of a convective thunderstorm 10,000 km northwest of Jupiter's Great Red Spot, obtained by NASA's Galileo spacecraft on June 26, 1996 (Credit: NASA/JPL-Caltech)

How the Planets Move

Galileo got to Jupiter by way of immensely clever mathematics. The laws of motion presented by Isaac Newton in his *Principia Mathematica* of 1687 were the pilots of our spacecraft. But Newton worked using orbital rules that he derived from Johannes Kepler. After Newton, these became enshrined as Kepler's Laws. We have already met Kepler the dreamer of the Moon. Now we can assess Kepler as a unique figure in the history of astronomy.

The two are not actually different. Kepler was an omnivorous thinker, devoting time to the study of snowflakes, Latin poems and astrology. He was a true humanist, converted to Copernicanism while studying at Tubingen between 1589 and 1594 under Michael Mästlin (1550–1631). Kepler's first astronomical work was a vigorous defense of the Copernican Solar System:

> As I listened attentively to the lectures of the famous Magister Michael Mästlin, I saw how awkward in so many ways the customary notion of the structure of the universe has become. I was delighted, therefore, by Copernicus....I have therefore bit by bit gathered together all the mathematical advantages that Copernicus has over Ptolemy.

This Copernicanism was not incompatible with religious faith. A devout Protestant believer, faith lay behind much of Kepler's scientific thought. Pondering the shape of the orbits of the planets, he perceived that an equilateral triangle could be placed between the orbits of Jupiter and Saturn. This may seem an odd idea to us, but by connecting planetary orbits with the perfect or Platonic shapes, Kepler believed that he was sensing the divine shape of the Solar System. The other Platonic solids that Kepler fitted into the observable Solar System were the tetrahedron, cube, octahedron, dodecahedron and icosahedron.

The data did not quite fit. But that was fine – the data must have been wrong, and Kepler knew where the best data was to be found. In January 1600 he met with Tycho Brahe, starting an awkward relationship that nevertheless produced great scientific fruit. Brahe's careful observations of Mars, which we have met already, formed a powerful record of the way the Solar System worked, and Kepler sought to make sense this by plotting an orbit for the planet. The circular orbit he tried first did not work.

This mattered. Everyone, Ptolemy and Copernicus alike, had assumed that circles described the orbits of the planets. Kepler measured the position of Mars relative to Earth every 687 days, when Mars had described a complete orbit, and found that Earth was not moving at a uniform speed and that the Sun was not at the center of the orbit. His words are worth repeating because they changed the way we understood the Solar System: "The planet's path is not a circle – it curves inwards on both sides and outward again at opposite ends….The orbit is not a circle, but an oval."

This conclusion ended a 2,000-year assumption that perfect circles dominated the heavens. Kepler's first two laws of planetary motion described the true state of affairs – that the orbits of the planets are ellipses, with the Sun at one focus, and that the line joining the planet to the Sun sweeps out equal areas in equal times as the planet travels around the ellipse.

It was not easy to make these discoveries known. But they did eventually emerge in Kepler's 1609 *Astronomia Nova*. His third law appeared 10 years later, and expresses how, for any two planets, the ratio of the cube of the mean distance from the Sun to the square of the period is the same.

Kepler continued to seek the deep secrets of the universe. In the *Rudolphine Tables* (1627) he put into effect his laws through a series of planetary tables. This was the consummation of Tycho Brahe's work, although publication was delayed because of squabbles with Tycho's heirs. Meanwhile Kepler searched on, his quest couched in religious language. He wrote in his *Harmonices Mundi* (1619) that: "When the full Sun illuminated my wonderful speculations, nothing holds me back. I yield freely to the sacred frenzy."

Delayed Light

Some 41 years after Kepler's death in 1630, the Royal Academy of Science in Paris mounted an expedition to Tycho's island in the Baltic Sea. Their purpose was part practical, to make measurements that might help fix longitude. But for one participant at least, the Danish astronomer Ole Römer (1644–1710) then working at the Academy, it must also have been a pilgrimage.

Römer had never met Tycho. But with his astronomy professor Erasmus Bartholinus, Römer had worked for 4 years preparing Brahe's observations for publication. The project had been canceled, but it was not a waste of time. It formed Römer's astronomical mind. He wrote:

> Having completely familiarized myself with Tycho's principles…I had in mind that part of astronomy which deals with the determination as accurate as possible of the positions and movements of the fixed stars and the planets. I never was of the opinion that the more prestigious parts of astronomy dealing with spots, moons, and the figures of planets are comparable to these.

Römer was not only a careful observational astronomer. He was also, like Tycho, a maker of instruments. In 1676, Römer demonstrated to the Paris Academy a telescope he had constructed with two eyepieces designed to measure apparent variations in the size of the Moon. A year later, he constructed a device that showed the motions of the Jupiter system – it was known as a Jovilabium.

Jupiter was important to Römer. He had made a strange discovery with the planet and its moons.

Since Galileo's time, successive astronomers had published tables predicting the orbits of the four Galilean moons. The best of these was Cassini's, but the details were still not accurate. Perplexed by the problem, Römer came up with a theory, suggesting in 1676 that the problem was not with the moons or their orbits but with the light itself. Observing carefully some 40 revolutions of the satellites, he concluded that the light itself was taking time to travel from the Jupiter system to Earth. Light was not instantaneous as Descartes had argued. It had a speed.

One might have thought that Römer's announcement would have heralded a new scientific revolution. It did not. The idea of light traveling at a finite speed was difficult to accept, and Römer did not advertise his achievement. The English translation of Römer's French paper covers a little less than two pages of the 1677 edition of *Philosophical Transactions*.

In addition, Römer had moved on. In 1681 he returned to Copenhagen. Appointed Chair of astronomy at the university, he re-equipped the Round Tower observatory and worked to prove the Copernican system by hunting out stellar parallax. He cut a slice out of the roof of his house to assist the arc of his telescopes. There is no record of his family's response to this, but it cannot have been too negative for it was on his father-in-law's estate that Römer constructed "the observatory that I have wished for 30 years." This, it seems, was what made him truly happy.

Today, though, Römer's discovery really does matter. The idea that what we see through our telescopes has taken time to reach our eyes is part of the wonder of astronomy, and we measure distances to deep sky objects in light years. When we look up at distant galaxies, we are observing history. When we turn our telescopes to Jupiter, we are repeating the observation where light speed was found. Just as Kepler's ellipses led to Newton's theory of gravity, so Römer's discovery of a finite speed of light led to Einsteinian physics and, as we shall see, to establishing the Hubble constant and the true size of the universe.

Römer also tells us something important about the astronomical method. Writing to the mathematician and philosopher Gottfried Wilhelm Leibnitz (1646–1716), who had urged Römer to publish more of his results in 1703, Römer explained: "I have the most mechanical of all the disciplines in the literary world (this is how I conceive of astronomy) that requires senses and hands to the same degree as thought."

"Mechanical" is not derogatory here. Römer uses a Greek word that carries with it a sense of care needing to be taken and of the value of repeated experiment. In words that will be familiar to many amateur astronomers, he points out: "There will always be something lacking or needing to be examined or adjusted, which cannot be added on the basis of speculations (this would be conceit) or achieved at a convenient moment: as a rule it requires tedious waiting."

Senses and hands to the same degree as thought. Römer was the true heir to Tycho Brahe. His quiet discovery of the speed of light was another element of the true shape of the universe, held in bright Jupiter and its jewel-like satellites. Galileo would have been so very proud.

Big Planet or Small Star?

Jupiter is one of the planets in orbit around the Sun. But the scientific understanding of Jupiter reveals basic differences between its composition and those of the rocky planets. Jupiter may have a solid core, but it is primarily a gaseous world, and in that sense is more akin to a star than to Mars or Earth.

There are groups of stars being discovered that blur the distinction between stars and planets. They are objects so cool that they fall off the end of the spectral scale of stars running from O to M. In 1995, the first discovery of a brown dwarf star was announced – Teide 1 in the Pleiades, named after the volcano on Tenerife from where the star was observed – and since then astronomers have created new spectral classes for these chilled and dim inhabitants of the night sky. L-dwarfs are so cool that dust condenses in their atmospheres, while T-dwarfs, with a temperature range of 500–1700 C, have blue methane atmospheres as found around Uranus and Neptune. T-dwarf stars are too cold to achieve hydrogen fusion, and so exist in a permanent stellar freeze. At time of writing, some 59 of these objects have been discovered.

There are other stars even cooler. Y-dwarfs have ammonia atmospheres and temperatures below 200 C. NASA's Wide-field Infrared Survey Explorer (WISE) mission has identified more than 100 Y-dwarfs in our own stellar neighborhood thus far. The coldest, WISE 1828 + 2650, is less than 25°. Studies of Spitzer Space Telescope data estimate that the Y-dwarf WD0806-661B, which vies with the WISE discovery for the coldest star in space, is just 6–9 times larger than Jupiter.

All of which raises the question – is Jupiter a planet or a failed star? Dynamically it remains in orbit and so is properly a planet, but in terms of composition we might see it more as a transitional object, caught between a rocky core and stellar aspirations.

Facts and Poetry

The sight of Jupiter through a telescope is breathtaking. Images such as this one, taken by NASA's New Horizons spacecraft on its way to the outer Solar System, only increase this sense of wonder.

Here is a striking quotation: "The facts of science are at least as full of poetry as the most poetical fancies."

Fig. 14.5 A plume from the Tvashtar volcano tops Io as New Horizons leaves the Jupiter system, taken on March 1, 2007 (Credit: NASA/JHU/APL)

These are the words not of a scientist but of Charles Dickens (1812–1870), an author sometimes presented as either ignorant or opposed to science. But when Dickens turned to write directly about science, undertaking a book review for a London magazine in 1848, he spoke of science's power to inspire.

His argument is interesting. He points out first of all that the science of his day had undermined many legends, from the existence of dryads and mermaids to astrology. Was this a loss to human imagination? Dickens suggests not: "Instead of binding us, as some would have it, in stern utilitarian chains, when science has freed us from a harmless superstition, she offers to our contemplation something better and more beautiful."

Science, the destroyer of shining cities at the bottom of the sea, had shown instead the real wonder of coral reefs, replacing the creatures of the deep underground with geology, "the great stone book which is the history of the Earth." As for astronomy: "Two astronomers, far apart, each looking from his solitary study up to the sky, observe, in a known star, a trembling which forewarns them of the coming of some unknown body through the realms of space."

This is not some astrological prediction but Dickens' account of the discovery of Neptune by John Couch Adams and Urbain Le Verrier in 1846.

Dickens' point is that the discoveries of science are a feast for the imagination, replacing myths with wonder. He tried to present these wonder-filled facts in his periodical *Household Words*, serializing the work of the great chemist Michael Faraday, an early example of popular science. Dickens would have agreed with Faraday that: "In the pursuit of physical science, the imagination should be taught to present the subject investigated in all possible, and even impossible, views."

How better to do this than by observing Jupiter?

Chapter 15

Meet the Gangs

Observation: Pleiades, Hyades, M44, M13, M92, Double Cluster
Significance: How galaxies form, expansion of the universe
Science: Globular and open clusters, Pop I and Pop II stars

All stars are born in clusters. Most escape from them. But stars still in their clusters, bound by gravity one to another, form some of the most distinctive observing targets in the night sky. In this chapter we will observe several of these, and also explore what makes a cluster and why clusters matter.

Astronomy distinguishes between two types of clusters. Open clusters are irregularly shaped and consist of a few hundred stars destined to separate over time. They are quite close to Earth and enable detailed observations of stars that formed at the same time and from the same material. Globular clusters are different creatures – they are densely populated, tightly packed balls of thousands of stars bound closely to each other, very hot and very old. Globular clusters take us back to the earliest aeons of galaxies, for they are the relics of how galaxies came to form. They have also played a central part in a story we have started but not yet concluded, the expansion of the night sky to the vertiginous vastness of the universe.

The Sisters

Our first observation takes us along a line that can be drawn between two of the brightest stars in the northern winter sky, Betelgeuse and Aldebaran. Continuing this line beyond Aldebaran leads to the beautiful naked-eye open cluster called the Pleiades or Seven Sisters in the constellation of Taurus the Bull. Under light-saturated skies, there are only six stars visible, but in dark conditions many more appear.

M. Marett-Crosby, *Twenty-Five Astronomical Observations That Changed the World: And How To Make Them Yourself*, The Patrick Moore Practical Astronomy Series, DOI 10.1007/978-1-4614-6800-4_15, © Springer Science+Business Media New York 2013

Fig. 15.1 The Pleiades, 20×3 min exposure at 800 ISO on Monday January 22, 2007 with a modified Canon 350D, and a 20-cm (8″) f/2.75 ASA Astrograph (Credit: Albert van Duin, www.astropix.nl)

If you want to identify which is which among the classical seven sisters, start with the bowl of the dipper-like asterism. These four, the brightest stars, are from west to east Maia, Electra, Merope and Alcyone. Asterope, Taygeta and Celaeno lie in a ring around their big sister Maia. The girls' parents lie to the east of the main group, a pair of stars called Atlas and Pleione. Alcyone, the brightest of the Pleiades stars, is sometimes called the Hen.

The Pleiades rewards the use of binoculars, which reveal the wide field of the stars. It is a strange alchemy of sight, but the brighter members, shining white and blue, seem to come staggeringly close to the eyes. They form a neat asterism not unlike that of the Big Dipper. Through larger aperture telescopes it is possible to detect very faint wisps of nebulosity around some of the stars. After observing these in 1859, the German astronomer Wilhelm Tempel (1821–1889) described them as "like breath on a mirror." One of these, NGC1432, seems to rest around the star Maia, while IC349 forms a shadow near the star Merope. However, comparisons of their relative velocities indicate that these areas of nebulosity, while lying in line of sight between ourselves and the cluster stars, are not in fact related to them but are outskirts of other dust clouds in Taurus and Auriga.

The brilliance of the Pleiades to the naked eye is not a lie. This is one of the closest clusters to Earth. The Hipparcos satellite measured it at some 390 light years away, but more recent studies have stretched the distance to 425 light years.

It is no distance in binoculars from the Pleiades to two other open clusters. Returning towards Aldebaran and placing that star in the center of a binocular field of view reveals the Hyades, apparently milling around Taurus' alpha star. This is another illusion created by line of sight – the Hyades have no relation to Aldebaran, the star being much closer than the cluster. That said, the Hyades is the closest open cluster to Earth, a mere 153 or so light years away. They also have no connection to the Pleiades save in myth – the name Hyades is, like Pleiades, derived from Greek legend, both the Pleiades and the Hyades being daughters of Atlas but by different wives. They are also very different through binoculars, the Hyades a pale yellow to the Pleiades' brilliant blue.

How old are these two clusters? The Hyades is some 625 million years old, with a margin of error of around 50 million years to either side. It is dominated by K-type giant stars. The Pleiades is younger, perhaps just 100–130 million years old – the figures vary – with 125 million years the most recent estimate for the cluster's age.

A third open cluster well worth observing in Taurus lies between the constellation's horns. Starting once again at Aldebaran, notice the bright star alongside it in the direction of the Pleiades. This is Epsilon Tauri or Ain. Forming the point of a triangle with these two stars as the lower line is NGC1647, a distant cluster some 1,660 light years from Earth and younger than the Hyades.

Labeled variously as Melotte 25, Caldwell 41 or Collinder 50, the Hyades are not in the Messier list. They are a moving group, part of a supercluster with a later target for our observing of the Praesepe or Beehive Cluster M44. It was Olin Eggen who identified this connection by studying the similar velocities of the stars and their shared iron to hydrogen ratios. The Pleiades are also part of a wider group known as the Local Association, which includes many stars within 100–150 parsecs of our own Sun.

(Olin Eggen (1919–1998) came to astronomy by way of being a pianist and wartime intelligence agent. He apparently also wrote some science fiction under the pseudonym Nilo Negge, not perhaps the hardest of codes to crack since it is simply his own name in reverse).

Almost all of the stars in the cluster lie along the curve of the main sequence. These are mostly B-type stars, although the cluster also includes a considerable number of brown dwarfs. Because they are reasonably near Earth, clusters are an important source of science for this kind of celestial object.

What will happen to these magnificent open clusters? When dealing in the scale of the lifetimes of stars, nothing is certain, but it seems that no open cluster can stay bound for much more than 100 million years. The gravity of other bodies can draw members away, while gravitational interactions between cluster stars can cause stars to be expelled. Meanwhile the effects of rotation around the galactic center will gradually stretch the stars apart. Models of evolution applied to the Pleiades suggest that its stars will be stripped away with increasing speed by tidal encounters with the galactic gravitational field. An observer looking at the stars in some 500 million years or so will see an elongated Pleiades with two outstretched arms extending from a core much smaller than that we can now observe.

Between Two Donkeys

It is not far from the Hyades to our next observing target, the Beehive or Praesepe cluster, M44. This is another naked-eye object under reasonably dark skies, although like the Pleiades it is best seen through binoculars. Galileo observed this cluster through his telescope in 1610.

The Beehive lies in the fairly indistinct constellation of Cancer the Crab. It's best found with reference to two stars outside Cancer, namely Pollux in Gemini and Regulus in Leo. Follow a line between these two bright stars stretching across the Northern Hemisphere's spring sky, and on a dark night you will make out a pair of stars a little less than half way towards Regulus, just below the imaginary line. These are Gamma and Delta Cancri, respectively Asellus Borealis and Asellus Australis, the Northern and Southern Donkeys when translated. Leaving aside the mystery of two donkeys in a crab, find the stars in binoculars, and they become a neat group of four stars with Eta and Theta Cancri. The cluster lies within the square they make.

Fig. 15.2 M44, the Beehive Cluster, with Saturn (Credit: Jimmy Westlake)

While familiar with this square of stars, use the two below M44 and move away from the cluster along the line they make. This points towards a third and then a fourth star, which is Zeta Cancri or Tegmen, the shell of the Crab. It is possible to separate Tegmen through a reasonable aperture, and at higher powers one of these further divides into a pair. The whole system contains at least four stars locked in orbits around each other.

M44 is a lovely observing target. Modern science identifies at least 400 stars in the cluster, and through an amateur telescope there are plenty of stars in view, although once again the whole sweep of the cluster is the real reward of this observation.

Like the earlier open clusters, M44 is not too far from Earth, some 610 light years. It is rich in stars exhibiting heavy elements in their spectra and includes a large number of variable stars. Their colors also tell a story. In contrast to the blue of the Pleiades, this cluster is flecked with the orange glow of K-type giant stars. It may share a common origin with the Hyades, and what we see as one open cluster may be an agglomeration of two. Studies have indicated that a sub-cluster with a mass some 30 times that of the Sun is merging with the main body, further indication of the dynamism of open clusters under the forces pressing on them.

M44 was first labeled as the Beehive Cluster during the nineteenth century. It possesses an earlier name, the Praesepe or Crib. It seems that the identification of the two nearby stars with donkeys by Greek observers – they were the donkey-mounts of two Greek gods on their way to battle giants – encouraged observers in the Christian centuries to see the straw-like object between the donkeys as the crib of the infant Jesus.

Great Globulars

Globular clusters make a fine northern summer observing activity as the constellation Hercules passes across the northern sky. Despite his grandeur this hero is far from obvious and is best found by way of Vega, the mighty star of the much smaller constellation Lyra. Identify Vega first and then Arcturus. On a line between them lie two very useful asterisms. The first is the squashed box known as the Keystone at the heart of Hercules, and the second, a little further towards Arcturus, is the curl of the Corona Borealis, the Northern Crown.

The Keystone quartet of stars are all of pretty much the same magnitude, and once you've seen them, they are easy to find again. The Keystone lies west of Vega. Don't drift north, as there is a deceivingly similar asterism in Draco. The best way to check is to follow the line between the pair of stars furthest from Vega. Through binoculars, it should contain one of the finest globular clusters that there is in the night sky, M13.

Take time to focus on Zeta Herculis, the star in the quartet furthest from Lyra. This is a double star of subtle colors, the primary yellow and the companion a delicate orange. They orbit each other every 34.5 years and are approaching a phase when separating them in an amateur telescope is a good observing challenge.

It is possible to spot M13 with the naked eye. Edmund Halley did in 1714, "when the sky is serene and the Moon absent." Through a reasonable aperture, the cluster reveals defined edges and a glowing core, the effect of some million or so stars jostling within a diameter of 160 light years. M13 is 26,000 light years from Earth.

Fig. 15.3 Hubble image of the nucleus of M13 (Credit: NASA/STScI)

A second globular cluster in Hercules can be found by returning to Lyra and then tracking from that star across Hercules above the line made by the upper pair of Keystone stars. M92 is a little less bright than M13 and substantially smaller than its neighbor. Under good conditions, binoculars are a good way to find the where-abouts of the cluster, but through a reasonably good telescope its shape and core become clear.

Averted vision is a useful trick to make the most of M92, and it's essential for a third globular cluster in Hercules, NGC6229. This lies outside the Keystone asterism, but use the two stars that form the edge furthest from Lyra and track upwards from the lower to the higher towards a fuzzy binocular patch. Spotted first by William Herschel who mistook it for a planetary nebula, it was also mistaken for a comet before settling into being identified as "a very crowded cluster" by Heinrich d'Arrest (1822–1875). This becomes a fine globular cluster under decent telescopic magnification, but it is a challenging find. At 100,000 light years from Earth, this is not too terribly surprising.

What are we looking at here? Globular clusters occur in all galaxies, and they are thought to be fossils from the very earliest period of galactic formation, some 13–14 billion years old in the case of the Milky Way. The number of such clusters varies from galaxy to galaxy, but the concentrations can be extraordinary. The M87 galaxy, which we will observe when we turn to the constellation Virgo, may contain as many as 13,450 globular clusters. The Milky Way seems paltry by comparison, with just 157 confirmed at time of writing, but as is often the case, face-on distant galaxies are easier to examine than the dust-laden realm of our home.

How are they formed? The traditional model has proposed that all stars in a globular cluster form at the same time as in open clusters, and measurements do confirm that some globular clusters came into being in this way. But others don't. Observations by the Hubble Space Telescope of NGC2808 in the constellation Carina suggest that, rather than a baby boom birth, this cluster was formed in three successive bursts of stellar generation. Even the ages of clusters are proving more complex than first thought. They are perhaps one of the celestial populations around which the most questions are gathering, as astronomers seek to probe their role in the earliest histories of galaxy formation and evolution.

Finding the Edge

With startling suddenness and definitiveness, they seem to have elucidated the whole sidereal structure.

So wrote Howard Shapley in a letter to Arthur Stanley Eddington in 1917 as he used the 60-in. Mount Wilson telescope to photograph Cepheid variables locked within globular clusters. It seemed to Shapley that they provided the evidence he needed to prove the size of the Milky Way, and it was with this that he entered the Great Debate, convinced that he had "a vague skeleton of the whole Galaxy" by virtue of his observations.

As we saw in an earlier chapter, Shapley's inflated Milky Way was a wrong reading of the stars. He had underestimated the degree to which dust in the interstellar medium dimmed the light coming from Cepheids, but he also assumed that all Cepheids behaved in the same way. It was this latter part of Shapley's model that was undone by Walter Baade (1893–1960), a German-born astronomer who used the 200-in. Hale Telescope at Palomar to photograph individual stars in the nucleus of M31. Baade had time. Classified an enemy alien during World War II, he was deemed unsuitable to contribute to the war effort in America. Los Angeles also suffered power losses, making for great dark sky observing.

So at the International Astronomical Union meeting of 1952, Baade was able to announce an error in the calibration of Cepheids. The effect was to double the size of the known universe and set in train the steady expansion of the distances we measure to other galaxies. He had already introduced in 1944 a distinction between Population I and Population II stars based on location, identifying in the Milky Way a central bulge and halo where the older stars lived, and a flattened disc divided into spiral arms where the younger stars were found.

This distinction between Pop I and Pop II remains, although since Baade's time the definitions have widened to include not only location but also composition and age, three interconnected elements that explain many of the observational results plotted on a Hertzsprung-Russell diagram. Pop I stars are young and metal-rich. Pop II stars are old and metal poor. And since it is in the disc rather than the bulge or halo that most of the unbound gas remains, it is in the disc that the young stars are being formed. The halo, meanwhile, harbors the fossilized remains of the original process by which the galaxy was formed. Observing globular clusters takes us to the beginnings of our Milky Way.

The Seated Queen

We have made use several times of the distinctive w-asterism of Cassiopeia, queenly mother of chained Andromeda and husband of Cepheus. He sits alongside her in the northern sky above Perseus, Andromeda's rescuer. The asterism is Cassiopeia's throne and is formed from five stars, four of them blue but one distinctively red. This is Alpha Cassiopeiae, the star Shedar, and it makes a fine contrast with the last star in the asterism and its near equal in luminosity, Beta Cassiopeiae or Caph. Caph is some 54 and Shedar 228 light years from Earth, making Shedar a star of huge intrinsic luminosity. Placing red Shedar in the center of a binocular field, move a little away from Caph to spot, south of the asterism, the emission nebula NGC281. This is worth trying to capture in a telescope for sure, although its shape in photographs, which has earned it the nickname of the Pacman Nebula, will not be obvious.

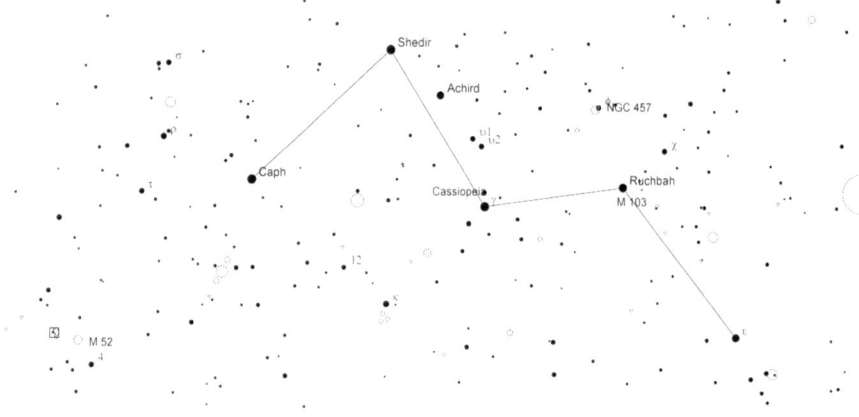

Fig. 15.4 The Constellation Cassiopeia (Cartes du Ciel)

Caph's full Arabic name is translated as 'the stained hand,' but this is not the site of a murder mystery. It refers to an interpretation of the Cassiopeia asterism as the five fingers of a hand. Why is the star stained? Because this finger is imagined as being marked with henna, a sign of beauty.

The hunt for clusters takes us back to Caph. Using binoculars, follow a line between the Alpha and Beta stars onwards for the same distance beyond Beta and you will be able to spot M52, one of the busiest open clusters in Charles Messier's list with some 130 stars in its main group. It is a fine binocular observation and has the additional pleasure of the red star 4 Cassiopeiae above it. Meanwhile just south of Caph, notice a pair of stars Ro and Sigma Cassiopeiae with, suspended in between them, the very striking open cluster NGC7789. With a telescope it is possible to penetrate quite deeply into this cluster, although a good pair of binoculars once again provides the richest overall view.

A Double Swarm of Stars

Cassiopeia's asterism points the way into Perseus. Moving along the shape of stars from Alpha Cassiopeiae away from Beta, the next star is Gamma and the one beyond Delta. In binoculars, get a sense of the distance between these two stars and follow the line they make two distances down beyond Delta. Poor seeing can sometimes make this process a bit frustrating, although they are naked-eye objects in the best conditions, but persevere because in this otherwise gloomy area of Perseus lies one of the treats of cluster observing, the so-called Double Cluster, Caldwell 14.

This has been a favorite object for astronomers since very early times. Hipparchus in the second century BCE refers to it as "a cloudy spot," and Garrett P Serviss, in a book delightfully entitled *Astronomy With An Opera Glass* (1888), proclaims it "a double swarm of stars." They form the handle of Perseus' magic sword and, like the sword of that other hero Orion, repay careful observation. The wide field is once again where the pleasure of this observation lies, but ensure you have enough power to make the separation.

The Double Cluster is made up of two bodies. The one closer to Cassiopeia is NGC869 (also known as h Persei), which persists in appearing to be nearer Earth although in fact it is the farther of the two. NGC884, or Chi Persei, has some noticeable red stars and an interesting red variable, RS Persei, at its center.

Fig. 15.5 The double cluster, imaged on November 6, 2004 at Landers, California, USA, with a ST-10XME/Vixen 102-ED/G-11 at f/6.5; exposures LRGB: L 8×2.5 min: R 8×2.5 min: G 8×2.5 min: B 8×2.5 min: RGB binned 2×2 (Credit: Alson Wong, www.alsonwongastro.com)

Are the two clusters related? They are both probably about 7,500 light years away – the precise distances are uncertain – and both contain some 300 stars. Both are also young, and it is straightforward to assess which of the two is older by the presence of some red stars in NGC884. It's likely that they are unconnected to each other and only seem to be neighbors because of where we stand to look, a view which may well get better over the next million or so years, for both of these clusters are blue-shifted and are approaching our Sun at about 14 miles a second.

Perseus is rich in clusters. The easiest to find is that around the constellation's Alpha star Mirfak. Put this star at the center of the binocular field, and you will gain a sense of the huge cluster Melotte 20, observed in an earlier chapter. It is a mere 50 million years old and forms part of a wider association of stars all with a common origin. Moving to the second brightest star in the constellation, our old friend Beta Persei, or Algol, see if you can find the open cluster M34, which makes a triangle with Mirfak and Algol.

There is also another double cluster in the constellation, a really challenging hunt near the +4 magnitude star Lambda Persei. The star lies on a line between

Mirfak and Capella, the bright alpha star of Auriga, and is part of a cluster of four stars, all of similar magnitude, in this rather indistinct area of the constellation. Just outside this cluster near Lambda Persei await the open cluster pair NGC1528 and 1545, some 2,500 light years from the Sun and so much closer to us than the more famous double. Again, it seems likely that there is no gravitational bond between the pair, but they make fine quarry for a nighttime hunt.

Sharing the Stars

From earliest times, humans observing the stars have wanted to record the Pleiades' beguiling patterns. They are a recognizable asterism on the Nebra Sky Disk, a bronze circle decorated in gold that was created by Nordic metalsmiths at some point during the European Bronze Age, between 1700 and 1600 BC. Aside from the Sun and the Moon, only the Pleiades are represented as they appear in the sky; all other stars are spaced evenly around the disk.

At something like the same time, the Babylonians were writing about them. They feature in a *Prayer to the Gods of the Night* composed between 1830 and 1530 BCE. In a quite different culture, the Skidi Pawnee Indians in North America made much use of the Pleiades as a means of marking the best time to start planting crops.

The seven visible stars of the Pleiades are also prominent in Australian Aboriginal legends. In mainland Australia, the stars were either seven sisters running as a group or six helping their youngest to escape from a man. The identity of that man varied, sometimes Orion but at other times the Moon, Venus or a star like Canopus or Aldebaran. There were footprints in rocks believed to have been created by their flight. Meanwhile in Arnhem Land, the stars were the partners of fishermen, their husbands represented by the stars of Orion.

It is striking that the Pleiades have so often been seen as female stars. It suggests tantalizing links across unconnected cultures. It also indicates, perhaps, that there is a common language of the stars that can transcend cultural specificity.

There are just a very few instances when we can glimpse how the stars can form bonds across cultural frontiers. In twelfth-century Spain, political conditions made possible an all too rare interaction between medieval Christian, Jewish and Islamic astronomers. They could read each other's languages and use each other's astronomical tables. Muslim astronomers such as Ibn al-Kammad (living around the year 1115) created tables that only survive because Latin speakers translated them, and a Hebrew translator states explicitly, "I depend for most matters" upon the work. The Latin and Hebrew works both retain the Muslim calendar and enabled users to find the positions of the planets.

We are told that King Alfonso X of Castile, who died in 1284, obtained a copy of an Arabic work of astronomy. When he had "understood the value and great profit which was in the book, he commanded him to translate it from Arabic into the Castilian language." Who was the him? Alfonso's Jewish physician-astrologer. Astronomy was a safe place wherein otherwise divided minds could mix.

Chapter 16

Life and Death

Observation: Ceres, Vesta, smaller asteroids
Significance: Conditions in the earliest Solar System, age of the Solar System, impact
Science: Asteroid belt, Kuiper Belt objects. meteorites

Our next observation takes us on a hunt for the smallest visible objects in our Solar System. Tiny they may be but also important. The cratered surfaces of the Moon, Mars and Mercury reveal the power they bear. In this chapter, we are going to start by exploring what we mean by asteroid, planet and dwarf planet. We will examine how they offer evidence for conditions in the very early Solar System and how they yet might change the world.

We have to lay aside with sadness one image of asteroids. In the second film of the first *Star Wars* trilogy, *The Empire Strikes Back*, the heroic ship Millennium Falcon escapes its pursuers by weaving through a close-packed, tumultuous asteroid field, complete with its own life form. It is necessary to report that nothing like this exists, so far as we know.

In the film, Darth Vader intones: "Asteroids do not concern me, Admiral." They do, though, concern us.

Asteroids, Planets and Dwarfs

In 1802, William Herschel read a paper to the Royal Society in London describing his observations of two newly discovered "moving stars." His interest in these stellar mysteries arose because their discoverers were not sure what they'd discovered. Giuseppe Piazzi (1746–1826) thought he had found a comet, and Heinrich Wilhelm

M. Marett-Crosby, *Twenty-Five Astronomical Observations That Changed the World: And How To Make Them Yourself*, The Patrick Moore Practical Astronomy Series, DOI 10.1007/978-1-4614-6800-4_16, © Springer Science+Business Media New York 2013

Olbers (1758–1840) was convinced he had seen a planet. "This curious object," as Herschel described Piazzi's find, was so puzzling that Herschel devoted time and ingenuity to achieving detailed observations, projecting images across his garden so as to measure relative sizes.

They were certainly very small. "We certainly cannot class them in the list of planets," he said, "for…Mercury, which is the smallest, if divided, would make up more than 135,000 such bodies." But they weren't comets, either. "We ought to distinguish them by a new name," Herschel said: "I shall…call them Asteroids; reserving to myself, however, the liberty of changing that name, if another, more expressive of their nature, should occur. These bodies will hold a middle rank, between the two species that were known before."

Piazzi named his discovery Ceres Ferdinandea, after the Roman god of agriculture and King Ferdinand III of Sicily. Olbers called his Pallas, a name from Greek mythology. They are now called (1) Ceres and (2) Pallas. All asteroids are formally identified in this way – the name, if there is one, is preceded by a number, sequential in order of discovery.

Herschel's term, asteroid, took many years to enter standard astronomical language, and the problem he raises has not gone away. Just what is a planet?

The answer advanced by ancient astronomers was based on motion. The planets were celestial objects wandering among the field of fixed stars. It was an answer that worked, allowing the number of planets to increase as more powerful telescopes extended the reach of our eyes through the Solar System. Thus when on February 18, 1930, Clyde Tombaugh (1906–1997) discovered a new, far-distant member of the Solar System, he called it a planet and Pluto joined the family.

This idea of the planet began to come unstuck after August 30, 1992, when David Jewitt and Jane Luu announced their discovery of an object in orbit beyond Neptune. It was the first of many. Now known as Trans-Neptunian Objects (TNOs) or Kuiper Belt Objects (KBOs), there are now well over a 1,000 identified, some larger than Pluto. Twenty-five were discovered in 2010 alone. Are these all planets? If not, then what is a planet after all?

The International Astronomical Union's Congress of 2007 met in Prague to find an answer. It was an evocative location. Prague was where Kepler had met Tycho Brahe in 1600. Their task was to define a planet, and there were two main approaches, one dynamic and based on motion around the Sun, the alternative physical, invoking a certain minimum size or particular shape.

The eventual resolution represented a compromise. A planet was defined as a celestial body in orbit around the Sun. Two further requirements were added: a planet had to be of sufficient mass to assume a nearly round shape and to have "cleared the neighborhood around its orbit." There had to be evidence of power. For those that failed this last test, there was the new category of *dwarf planet*. Pluto and Eris were the first to be so designated, Haumea and Makemake following a little later.

All of this made sense within the IAU's rules for nomenclature. But it did not meet with overwhelming public approval. Owen Gingerich, chair of the Planet

Definition Committee, reflected that: "In their zeal for science, the voting astronomers in Prague seemed to forget that for the most part they don't own the telescopes.... It is the taxpayers who own them.... I realize in retrospect that the IAU should never have attempted to define the word planet. It is too culturally bound."

A planet, then, is not solely an astronomical phenomenon. Losing Pluto made some people feel deprived of something that they thought to have possessed. It might be that there are some things with too strong a hold on the imagination to be dismissed by redefinition.

Hunting Asteroids

Observing asteroids and dwarf planets is about hunters and prey. Asteroids are small, elusive and always on the move.

Fig. 16.1 Images obtained nearly simultaneously in the morning of September 29, at 02:30 h UT, when the asteroid was passing through the constellation of Triangulum Australe, The Southern Triangle. The offset between the two trails corresponds to the difference of the lines-of-sight from the two telescopes towards the object. Two 1-min images were taken almost simultaneously with the FORS-1 instrument on Kueyen, the second 8.2 m VLT Unit Telescope on Paranal, and on the WFI camera installed on the ESO/MPI 2.2 m telescope at La Silla (Credit: ESO)

This image shows asteroid (4179) Toutatis passing through the constellation of Triangulum Australe (the Southern Triangle), as captured by two large terrestrial telescopes. The evidence of the asteroid is its trail across the star field. Because this asteroid was passing within a million miles of Earth, that trail is relatively long despite the short exposure time. Toutatis is an atypical asteroid in that its orbit involves such close approaches to Earth and Mars –it is one of the largest known 'Potentially Hazardous Asteroids,' but in other ways it shows some of the characteristics of the asteroid family. It's fast and, up close, a spinning, non-circular body.

The two brightest asteroids in the night sky – (1) Ceres and (4) Vesta – are theoretically naked-eye objects, but to see them without help requires fabulously good eyesight and perfect dark skies. Normally, asteroid hunting requires at least a 6-in. telescope and a clear sense of where they are.

Asteroids move quickly against the background of the stars, so to identify an asteroid, you need to have a good idea of how the star field will appear. Astronomy magazines include sections that lead observers to dwarf planets in the main Asteroid Belt; there are also online resources. You may well find that you need the star map beside you as you hunt, so the magazine page or a computer printout are ideal. Alternatively, you can take out a laptop that is protected against getting damp with dew and use a sky search program. There are several designed with asteroid-hunting in mind.

(1) Ceres and (4) Vesta are normally some distance from each other in the sky, but their orbits do sometimes bring them into the range of a single observation. In 2012, for example, both of them were visible in Taurus. Look out for any occasion when the asteroids are at opposition. Ceres can be as bright as magnitude +6.7, bringing it comfortably within range of binoculars. It is normally in the range of +7.5 to +9.5.

Vesta can be more generous with its magnitude, as much as +5.3 at the most favorable oppositions. In such conditions, the asteroid is comfortably within range of binoculars on moonless nights, passing just over an astronomical unit from us. It is also unusual among asteroids in being highly reflective, bouncing back 26 % of the Sun's light towards the observer.

Because of the importance of accurate information when hunting asteroids, Internet resources are almost essential. Here are some:

- http://minorplanetcenter.net/iau/mpc.html – this is the homepage of the Minor Planet Center.
- http://ssd.jpl.nasa.gov/sbwobs.cgi – a NASA/JPL tool that will tell you what small bodies are available at the observation time and location you want.
- http://asteroid.lowell.edu/ – lots of services provided by the Lowell Observatory. You can do almost everything with the tools they provide.
- http://newton.dm.unipi.it/neodys/index.php?pc=0 – provides excellent information on near-Earth objects (NEOs).

You can seek to match what you're seeing to background stars by creating a bespoke asterism. This was how Vesta was uncovered from the star field by its discoverer, Heinrich Olbers, some 5 years after he identified (2) Pallas.

What will (1) Ceres look like? It is the largest of the main belt asteroids, and the images shown here were acquired from the Hubble Space Telescope. Even they are fuzzy. For amateurs, asteroids will always appear as points of light.

Ceres has an element named in its honor. Cerium was first identified in 1803–4, soon after the asteroid was spotted. Its oxide is used to polish high quality glass, including astronomical lenses.

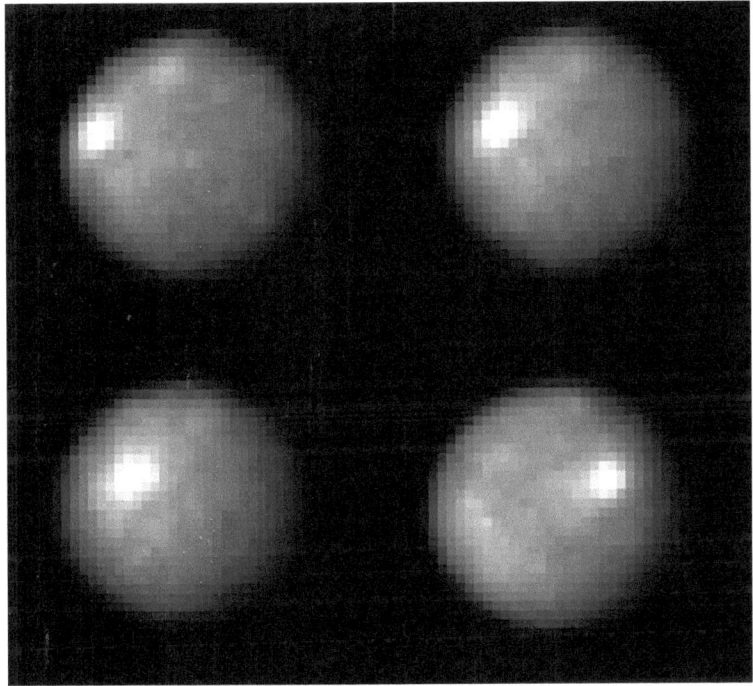

Fig. 16.2 Rotation of Ceres (Credit: NASA, ESA, J. Parker (Southwest Research Institute), P. Thomas (Cornell University), and L. McFadden (University of Maryland, College Park))

For the smaller asteroids, it is essential to have accurate data. This is best obtained via the Harvard-Smithsonian Center for Astrophysics' Minor Planet Center, Minor Planet Center for short. But even with good information, it is hard to be certain on any one observing night that you have really seen an asteroid. Try to imagine that your suspect belongs in a triangular shape with two stars and make a couple of sketches. Then look again later. If the object you thought was an asteroid is exactly where you last saw it, and the triangle is the same shape, then it was not an asteroid. But if it's moved, then you have your eureka moment.

This observing method is actually how professional sky surveys work, albeit with a more sophisticated kit. It is also how Clive Tombaugh discovered Pluto. He took two photographic prints of an area of Gemini divided by a gap of several hours. He then compared the plates and identified an object that had moved. This process is automated now. Digital images are acquired and then superimposed so as to identify the inconstant as opposed to the comparatively stable stars.

The first asteroid to be identified by looking at pictures was (323) Brucia in 1891, the first of many identified by the German astronomer Max Wolf (1863–1932). It is named after Catherine Wolfe Bruce (1816–1900), a philanthropist who gave generously to many observatories.

Is it possible for an amateur to discover a new asteroid? The answer to the question is yes, but it is painstaking work and requires serious equipment. With so many hyper-accurate professional sky surveys working through the data, amateurs need to be able to spot objects of magnitudes above +16 if they are going to have any chance of finding something new, but magnitudes of +20 are where most of the unknowns are thought to hide. The Minor Planets Center website includes a system for reporting new finds, and there are online groups dedicated to achieving these observations.

There is another way for amateurs to make a contribution in this area. The Lowell Observatory asteroid service publishes a list of asteroids in need of observation. Yet another way to make a difference to the science of asteroids is to seek out occultations, the passage of asteroids in front of stars or other celestial objects. Multiple observations can uncover details of the asteroid's shape and orbit.

Rubble Science

Why does spotting asteroids matter? Along with comets, they are the most numerous bodies in the Solar System. They are a challenge to observe and yield their secrets only to the most determined. But perceiving and understanding asteroids affects our understanding of our past and future.

Why? Asteroids are relics from the earliest ages of the Solar System. Very few of them have differentiated interiors. They are bits of what we used to be. There is reason also to believe that asteroids may be involved in the transport of minerals and even life across the empty wastes of space. Finally, asteroids hold our future in their unstable grasp. A large asteroid impacting with Earth will matter very much.

Let's start by establishing the where and what of asteroids. (1) Ceres and (4) Vesta inhabit an orbit between Mars and Jupiter some 2–4 AU from the Sun in the main Asteroid Belt. This zone puzzled the great astronomers of history, who felt that the gap should have contained something substantial. "Between Jupiter and Mars I place a planet," wrote Kepler in his *Mysterium Cosmographicum* (1596), "invisible on account of its tiny size." It was while searching for this missing planet that Piazzi discovered Ceres.

It made sense after Piazzi and Herschel to view the inhabitants of the main Asteroid Belt as the remains of a shattered planet. Now, however, asteroids are thought not to be shards of a broken world but the raw material for making one. The belt contains a multitude of fragments that never accreted into a single body.

Speaking in 1866, American astronomer Daniel Kirkwood (1814–1895) said that "the number of cometoids thus encountered, in the form of meteoric stones, fireballs, and shooting stars, in the course of a single year, amounts to many millions." He established from measurements of those asteroids whose periods were known at that time and, from this data, the existence of "gaps or chasms…analogous to those in the ring of Saturn."

What stopped them from combining? In 1761, the Swiss astronomer and mathematician Johann Heinrich Lambert (1728–1777) asked, "Are Jupiter and Saturn destined to plunder forever?" He had perceived the power exercised by Jupiter in this region of the Solar System. The gas giant, stirring up the asteroid population with its gravity, has kept them apart. The proof of this influence is revealed in the existence of gaps in the main belt, named Kirkwood Gaps after Daniel Kirkwood. These gaps correspond to orbital resonances with Jupiter.

There are less populous groups of asteroids outside the main belt. In order of distance from the Sun, these begin with Vulcanoids, asteroids within the orbit of Mercury. The next group out are the near-Earth asteroids, subdivided according to their orbit into the Apollo, Amor and Aten families. Trojan asteroids occupy stable orbits in the Lagrange points in the orbits of the planets Mars, Jupiter and Neptune and also Earth. On July 27, 2011, NASA confirmed the existence of Earth's first Trojan, 2010TK$_7$, with a magnitude of 20.7 and a diameter of around 300 m. It is likely to escape our planet as gravitational perturbances and other effects come to bear upon it.

Beyond Jupiter lie two further groups of asteroids. Centaurs occupy the region between Jupiter and Neptune, and they deserve their label. Part asteroid but also part comet with small comas of gas, their mythological ancestors were similarly divided, half horse and half human. There is no clear theory as to how Centaurs come to occupy their orbits.

At their outer range, Centaurs merge with the much more numerous Trans-Neptunian or Kuiper Belt Objects. They occupy a vast slice of the outer Solar System. Pluto and Eris are the largest thus far observed, but other sizeable objects include Makemake, Haumea, Sedna and Orcus. Some exist in systems. Pluto has three named moons, Charon, Nix and Hydra, a fourth which currently bears the designation S/2011 P1 and now a fifth, identified from Hubble Space Telescope images taken in June and July 2012.

Fig. 16.3 Mosaic of two images of asteroid (951) Gaspra taken by the Galileo spacecraft from a range of 5,300 km (3,300 miles), some 10 min before closest approach on October 29, 1991 (Credit: NASA/JPL)

Main belt asteroids come in many shapes and sizes. Those classified as dwarf planets are broadly spherical, but most asteroids are irregular, even improbable-looking objects. (951) Gaspra, the asteroid shown here in a close-up image taken by the Galileo spacecraft, turned out to be elongated and cratered, perhaps snapped off from a larger body by a huge impact. It has continued to undergo bombardment since. Such non-spherical asteroids are, when less than 150 m or so in diameter, solid bodies or monoliths. Larger asteroids are fragments bound together. Only the very largest are differentiated in the manner of the terrestrial planets into regolith, rocky mantle and iron core.

Some asteroids are not alone. Approximately 30 % of Kuiper Belt Objects and 2 % of main belt asteroids belong in binary systems, and it is thought that these pairs may be a source of some of the smaller satellites of the giant planets, many of which have highly elliptical retrograde orbits. Triton, the principal satellite of Neptune, may have been plucked out of a binary that had strayed from the Kuiper Belt.

Asteroids are often classified by the manner in which they reflect sunlight. This can reveal their composition and can indicate specific surface features.

The first asteroid we came to know well was (433) Eros. The NASA mission NEAR-Shoemaker flew close to it in 1998 and entered orbit in 2000. Eros belongs to the near-Earth asteroid group, and a desire to learn about these potentially hazardous objects led to an unexpected achievement when on February 12, 2001, the probe landed on the asteroid. It obtained 69 close images before touchdown and survived the impact, despite never being designed as a lander. NEAR was able to reveal a boulder-strewn terrain, with fine dust gathered in smoother areas called 'ponds.'

An asteroid such as Eros is classed as an S-type, its appearance consistent with a 'stony' composition. Asteroids exhibiting a gray color are C-types, indicating a composition rich in carbon compounds and other dark materials. They tend to be more common in the outer part of the asteroid belt.

NEAR-Shoemaker showed the way for other close encounters. The Japanese Hayabusa probe landed on (25143) Itokawa on November 20, 2005, and achieved the spectacular feat of bringing dust particles from the surface back to Earth. NASA's Dawn mission is currently journeying between Vesta and Ceres, seeking to compare a differentiated sphere (Ceres) with its potato-shaped cousin.

NASA's Dawn mission has revealed (4) Vesta in unparalleled detail, and the implications of its observations are still being assessed. Among many fascinating details, Dawn has been able to image the impact basins at the asteroid's south pole, now named Rheasilvia (formed around one billion years ago) and Veneneia (about a billion years before that). The basic crater structures are familiar, but Rheasilvia includes spiral deformation patterns not seen elsewhere. It is thought that ejecta from these two massive strikes may be the source of a group of meteorites known as HEDs that have been important in establishing the chronology of the early Solar System. Meanwhile Dawn's observations remain to be fully explained, for example the equatorial troughs previously unknown in the Solar System.

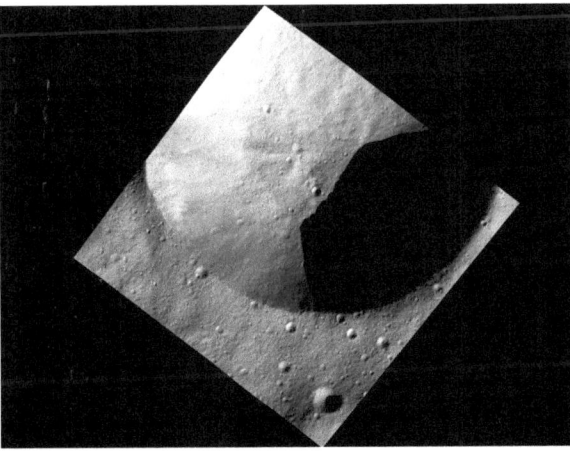

Fig. 16.4 The image, one of the first obtained by NASA's Dawn spacecraft in its low altitude mapping orbit, was taken by Dawn's framing camera on 13 December 2011 and shows part of the rim of a fresh crater on the giant asteroid Vesta (Credit: NASA/JPL-Caltech/UCLA/MPS/DLR/IDA)

We can approach asteroids even more intimately by studying fragments that have fallen to Earth as meteorites.

A brief word on the words. Fragments of asteroid hurtling through space are called meteoroids. They only become meteorites when they land on Earth. Most small meteoroids don't make it but burn out in the atmosphere as meteors.

A scientific collection of meteorites is an extraordinary thing to see. Lumps of rock scorched by their descent through our atmosphere, they are time capsules. The decay of radioactive isotopes in these fragments makes it is possible to date our Solar System. There is disagreement as to the absolute ages of a few components, with some scientists preferring 4.571 and others 4.566 as the age when the most primitive asteroid pieces started to form. But we are able to say with confidence that our Solar System is close to 4.56 billion years old.

Asteroids may also be a source of water. Some meteorites contain hydroxyl, and there is evidence of water ice in the spectroscopy of the asteroid (24) Themis. Themis is too hot for frost to form on its surface, so the water is likely to be a relic of the early Solar System protected from the Sun by a blanket of rubble. It is possible that an asteroid impact on the young Earth brought water to the planet, a gift from the wild heavens.

Did they bring life as well? During the Apollo era, groups of single-celled organisms, plant seeds and animal eggs were sent to the Moon so as to investigate the effects of the deep space environment on survival. The animal eggs especially did not do well. But later experiments on the space shuttle suggested that some extremophile life forms could survive in the vacuum of space. We will look at these experiments in more detail later. They were among the organisms included in the LIFE package, designed to travel to Mars and Phobos and then return to Earth on the Russian Phobos-Grunt mission of November 2011. It was a test of a theory called transpermia, which postulates life spreading between terrestrial bodies using asteroids as a form of transport. But the LIFE package died with the spacecraft when Phobos-Grunt failed to leave Earth orbit. We don't yet know if transpermia is possible.

Asteroids are not all gift. Increasing attention is being devoted to the threat they pose. This is the final reason why observing asteroids may yet change human history, substantially and forever.

Fig. 16.5 The barringer crater

It has happened before. Earth's surface bears witness to large meteorite strikes. This picture is not of the Moon but the Barringer Crater in Arizona. It is thought possible that a much larger impact caused the mass extinction at the Cretacious-Tertiary boundary in Earth's geological past.

Astronomers now distinguish between the bulk of near-Earth objects and the smaller number of potentially hazardous asteroids (PHAs), with the Torino Impact Hazard Scale defining the probability of impact and the damage, ranging from 0 damage to global catastrophe at 10. Identifying PHA asteroids involved detection and then confirmation of orbit via the Minor Planet Center's NEO Confirmation Page at www.minorplanetcentre.net/iau/neo/ToConfirm.html.

The first celestial object to be identified in this way was $2008TC_3$, imaged here by Peter Birtwhistle.

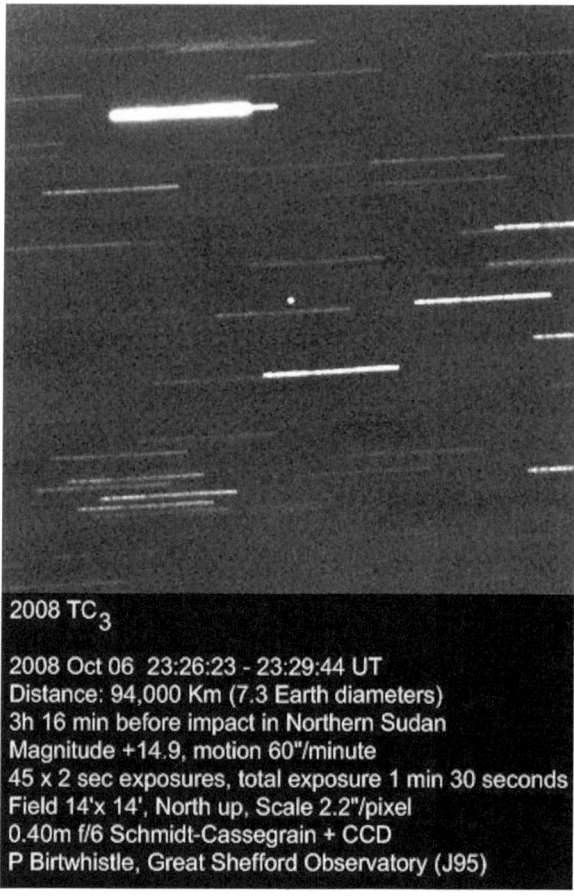

Fig. 16.6 Details on image (Credit: Peter Birtwhistle, www.birtwhistle.org)

In the end, this particular asteroid posed no threat, but how would we respond if one did? The obvious response is to attempt to destroy it, but this would probably create a swarm of small projectiles that might hit Earth anyway, all of them contaminated by the initial nuclear missile strike. So an alternative approach is to deflect incoming asteroids into a non-threatening orbit.

One asteroid where these systems might have to be tested is (99942) Apophis. According to current measurements, it will pass close to the planet in 2029 and again in 2036. Apophis weighs in at some 46 million tons. That is heavy enough to change Earth a lot and forever.

Don Quixote Rides Again?

If Earth is to be saved from the ultimate threat of destruction, it might be by a spacecraft called Don Quixote.

This is the name given by the European Space Agency to an ambitious two-spacecraft project that may be used to divert incoming threats such as Apophis. One spacecraft in the pair is Sancho, an orbiter that will gather information about the asteroid's composition. To get this kind of information, Sancho will need to resolve the surface of the asteroid at meter height and measure the Yarkovsky effect, the way the asteroid is emitting heat. This will reveal what may happen to the asteroid after it gets hit.

The second spacecraft in the pair is Hidalgo, the impactor. It will be called out of a parking orbit to change the asteroid's semi-major axis – to bash it off orbit, in non-space speak. If Hidalgo strikes while the asteroid is passing through a 'keyhole,' a particular moment in its orbit, then a change in axis of 100 m would be enough to render it safe.

Why Don Quixote, Sancho and Hidalgo? The names come from the novel *Don Quixote* (1605) by Miguel de Cervantes. Don Quixote is the befuddled but idealistic knight-errant who sets off on a heroic quest across La Mancha in Spain. Sancho Panza is his squire and Hidalgo means "knight."

Culturally, Don Quixote has meant different things, from Romantic hero to an anachronistic madman. It is from his name that we get our adjective *quixotic*, meaning both chivalric and romantic but also unrealistic, doomed to fail. Is it the quixotic nature of the project that inspired the name?

Or perhaps the whole mission might be seen as one worthy of a great knight, an act of love, as it were, with Earth as the new Dulcinea, the woman in the novel who is the romantic inspiration for the knight's adventures. In an English-speaking context, the name also carries echoes of Graham Greene's re-working of the Don Quixote story. He made the knight into a saintly figure as Monsignor Quixote.

But Cervantes' Don Quixote is most famous for tilting at windmills. The delusional knight thought they were giants with flailing arms. So perhaps the name captures the mission's ambivalence. We don't know whether the threat from near-Earth asteroids is great. Nor can we be sure that our response will work.

However, Don Quixote may yet ride again. He may have to, to save us all.

Chapter 17

Science, Fear
and Supernovae

Observation: The Crab and Veil nebulae
Significance: The deaths of stars, new stars' effects on humans
Science: Cataclysmic variables, novae and supernovae, SN1987A, pulsars

"When I had satisfied myself that no star of that kind had ever shone forth before, I was led into such perplexity by the unbelievability of the thing that I began to doubt the faith of my own eyes."

So wrote Tycho Brahe, recalling his experience of seeing a new star in the constellation Cassiopeia during a pre-dinner walk on November 11, 1572. Fortunately, he was not alone. Turning to the servants who were strolling alongside him, the great astronomer asked them to look as well: "They immediately replied with one voice that they saw it completely and that it was extremely bright."

Their reassurance allowed Tycho to believe that there was indeed a new star in the sky. He wrote a book about it, and its title gave astronomy the label "nova" to describe sudden additions to the city of the stars.

Tycho's book was titled *De Nova et Nullius Aevi Memoria Prius Visa Stella*, translated as "On The New Star, Never Before Seen in This Life Or in the Memory of Anyone." The Latin *nova* is the adjective *new*. The plural form for new star is *novae stellae*, so the plural in English for nova or supernova is novae and supernovae.

This chapter is about new stars, both novae and supernovae. Catching sight of sudden eruptions on distant stars is difficult, and our main observations will therefore be of remnants, the strange and beautiful fields of material left behind by these cataclysms.

Amateurs observe supernovae all the time. Their appearances are unpredictable, but when a new star is announced, telescopes swing through the sky to try to find it.

M. Marett-Crosby, *Twenty-Five Astronomical Observations That Changed the World: And How To Make Them Yourself*, The Patrick Moore Practical Astronomy Series, DOI 10.1007/978-1-4614-6800-4_17, © Springer Science+Business Media New York 2013

Fig. 17.1 SN2012 aw imaged on 27th March 2012 using an IR-modified EOS 450D camera attached to a Hyperstar convertor lens on a Nexstar 11GPS scope (10×45 s f/2 images stacked then processed with PS CS4) (Credit: Eric Walker)

This picture shows what success looks like. The blue arrow marks SN2012AW, a previously unremarked star that burst into brightness near the M95 galaxy in Leo about 38 million light years away. The explosion itself took place a very long time ago. The light from the star reached Earth on March 16, 2012. This image, taken from the Highlands of Scotland on March 27, shows the supernova in its galaxy and also M96, another member of the Leo I galaxy group. Another supernova was discovered in that galaxy by an amateur astronomer in 1998.

M95, together with M96, M105 and NGC3384, lies below the body of Leo. Identify the four stars that form the body using the two above Regulus in the backwards question mark asterism, and then search below the lower line of the body using binoculars. Don't get too near Regulus but look for the +3.8 magnitude Rho Leonis. The M96 galaxy will probably be visible first. Through a small telescope, the group of four will appear in a single field of view. They are some 38 million light years away and are the bright quartet in a larger group of galaxies.

Bright new stars like this one have mattered a lot to humanity. They were meticulously recorded as 'guest stars' in Chinese annals and painted onto the walls of caves by Native Americans. Their appearance has caused fear and wonder because of what they might or might not portend. One the most powerful and humbling discoveries science has made about the stars is that they live and die without reference to us at all.

New and Super-New

It was the science fiction author Robert Heinlein (1907–1988) who first wrote of an object, "It's been novaed." Since then, the idea has entered the popular mind such that most people know that stars explode. But Heinlein should properly have said, "It's been supernovaed" – a nova and a supernova are not the same.

One they were. Tycho Brahe's label 'nova' was applied for many centuries to any new bright star. It was not until the middle of the twentieth century that supernovae and novae were distinguished. Put simply, a supernova is the flash of a dying star. A nova star exhibits sudden changes in brightness but it continues to live.

In an earlier chapter, we noted the existence of a class of intrinsic variable stars called cataclysmic variables. Both novae and supernovae belong in this group.

Novae are classified into dwarf novae, the gentlest of the cataclysmic variables; recurrent novae, which flare every ten to a hundred years; and classical novae, where the change in brightness is greatest.

A nova may brighten by as much as ten magnitudes in a few days, followed by a gentle decline towards its original luminosity. Novae are usually associated with white dwarfs in binary systems with main sequence stars, and the outbursts are connected to the transfer of mass between the pair. Stellar material is flowing from the larger star towards an accretion disc and then to the white dwarf. Flares occur as this process becomes unstable. Although the ultimate fate of these white dwarf binary systems is uncertain – they may fizzle out as the matter stabilizes or they may explode – the nova event is part of the life of the system. It does not mark its end.

The most straightforward recurrent nova to observe lies in the constellation of the northern crown, the star T Corona Borealis. The Blaze Star, as it is sometimes called, is normally a +10 magnitude star, but it flared to magnitudes of 2–3 in 1866 and again in 1946; it could do so again.

The Blaze Star lies just outside the ring of stars that gives this constellation its name. Starting at the brightest star in the circlet (Alpha Coronae Borealis, named Alphecca or sometimes Gemma), move three stars around the ring in the direction away from Arcturus, the very bright star that presides over neighboring Boötes. This third star is Eta Coronae Borealis. Just outside the ring near Eta lies the Blaze. Keep watching, and who knows what you might see.

The same guide star Eta Coronae Borealis also points towards one of the strangest cataclysmic variables in the sky. Sometimes called a reverse nova for its strange habit of dimming, R Coronae Borealis is the prototype for the small and strange RCB group of variables. First observed by the British amateur astronomer Edward Pigott (1753–1825), this sixth magnitude star plunges into gloom by as much as nine magnitudes every few years.

Why? The mechanisms of RCB stars are not fully understood, but these hydrogen deficient stars are highly enriched in carbon and are probably cloaked periodically by carbon soot. At time of this writing R Coronae Borealis is beginning to recover from an episode. It can be found pretty much the same distance within the constellation's ring as T Coronae Borealis lies outside it, with our guide star Eta in the middle. It is worth trying to spot this unusual and interesting star, whose place in the story of stellar evolution is not yet clear.

On February 23, 1987, a night assistant at the Las Campanas Observatory on the Andean peaks of Chile noticed a star changing in the night sky. An amateur astronomer from New Zealand, Albert Jones observed it as it rose to magnitude 4.3, paused there for a time and then brightened to 2.9, becoming a naked-eye object.

Albert Jones described the discovery in an interview with Mike Simonsen published on simostronomy.blogspot.co.uk: "I was monitoring some stars not far from the Tarantula Nebula. On that fateful night, while I was observing stars elsewhere in the sky, I noticed some clouds coming over so I poked the telescope at my targets in the LMC. I was quite surprised to see a bright stranger, so I noted its position on the chart. But before I could make a magnitude estimate the clouds moved over.... Then the clouds moved away and I made an estimate of the stranger before phoning."

SN1987A was a test of the science of supernovae. It was remarkably suited to the task. It appeared within the Large Magellanic Cloud, a satellite galaxy to the Milky Way, comparatively close to Earth, at a known distance and with little dust to mask its struggles. It was also a star that had been observed carefully in the past, cataloged by the American astronomer Nicholas Sanduleak (1933–1990) in his survey of the Large Magellanic Cloud as SK-69 202.

SK-69 202 had been nothing very special, a blue supergiant some 20 times larger than the Sun near the Tarantula Nebula. So what had happened to it?

The hydrogen powering a star is a vast but finite resource, and as it runs out the central pressure of the star drops. This has the effect of raising the internal temperature due to compression. The result is the beginning of helium fusion and the subsequent creation of higher elements, the core temperature rising with each move through the Periodic Table. After helium comes oxygen and carbon fusion and then other elements towards iron. Once the iron core has grown to more than 1.4 times the mass of the Sun, beyond the Chandrasekhar limit, further shrinkage of the core and huge rises in temperature set in motion a chain of cataclysmic and rapid series of collapses. In the cool words of a review article concerning the death of SK-69 202, "In one second a configuration the size of Earth collapses to one with a radius of only 50 km. The velocity during the collapse reaches about 70,000 km per second in the outer portion of the iron core."

Although the internal processes involved are not fully clear, the consequences are. SK-69 202's core collapsed – perhaps into a neutron star, although that has not been detected – while the outer layers were torn apart in a gigantic explosion, releasing massive amounts energy. Most of this energy escaped as neutrinos, 10^{57} of them, released into space.

Of all those countless billions upon billions, only 19 neutrinos interacted with three detectors on Earth – at the Kamioka experiment in Japan, in a salt mine in

Cleveland, Ohio, and deep within the Caucasus in what was then the Soviet Union. But the 19 were enough. Along with the visual evidence, SK-69 202, which was now SN1987A (or had become some 160,000 years earlier, but we were only just seeing it), confirmed that the theory of supernovae was valid.

SN1987A was a Type II supernova. The distinction between Type I and Type II is made according to the presence or absence of hydrogen lines in their spectra. Type II show these lines and Type I do not. There are then further classifications within each type. It is now argued that Type Ia supernovae occur when white dwarf stars explode and Type II when supergiants reach the violent ends of their lives.

At time of this writing, NASA's WISE mission has identified a star that has brightened one hundred times since it was photographed in 1983. J180956.27-330500.2, as the star is labeled, is a red giant that has recently erupted into space, producing a cloud of dust equivalent in mass to Earth.

I Discovered a Nebula!

Are these the happiest words an astronomer can write? They were penned by Charles Messier on August 28, 1758. He continued, "This nebula resembled the Comet in the form, the light and the size."

Messier was looking for comets, not deep sky objects. It was in order to separate the wheat from the chaff that he started to put together his list of similar but confusing smudges in the sky, "So that astronomers would not confuse these same nebulae with comets just beginning to shine."

The first of these was in the constellation Taurus the Bull. It carries the label M1, and, since Lord Rosse penned his description, it has been called the Crab Nebula. It is the only supernova remnant among Messier's company of deep sky wonders.

There are two ways to find the Crab. The first is to begin at Aldebaran, the alpha star of Taurus. If you follow the line of Orion's belt from left to right it is an obviously brilliant star, one of the brightest in the night sky notwithstanding the meaning of its name – Aldebaran in Arabic denotes "follower," the star in pursuit of the seven sisters in the Pleiades. Aldebaran is an interesting star that has slid off the main Hertzsprung Russell sequence with the end of hydrogen fusion and is expanding as helium fusion begins. From Aldebaran, move towards Castor and Pollux, and in binoculars you will be able to identify the guide star Zeta Tauri above Orion.

Alternatively, start at Betelgeuse. Betelgeuse and Aldebaran form a fairly obvious triangle with a third star above them, the bright star Elnath at the top of Taurus. If you follow with binoculars a line between Betelgeuse and Elnath, there is only one bright star and that is Zeta Tauri. The Crab Nebula is just alongside the star, a little further in the direction of Elnath.

Elnath, sometimes Alnath or Al-Nath, has two Bayer designations. It is Beta Tauri and also Gamma Aurigae. It is almost always known by the former. Its name

Fig. 17.2 M1, the Crab Nebula, imaged with a Ritchey-Chretien 32" Telescope at f7.2 (prime focus) and SBIG STL-11000 CCD camera on 17 November 2004 (Credit: Jim Misti, http://www.mistisoftware.com)

means "the butting one" – it forms one of the horns of Taurus – and at magnitude −1.65 is one of the sky's brighter stars. It is also only a few degrees off being the point directly opposite the center of the galaxy.

The Crab is not the most immediately appealing of the Messier objects. Through a moderate amateur 'scope it can be a disappointing smudge, and it shows a remarkable ability to disappear in moonlight. At magnitude +8, it is the most challenging of the early Messier targets, and to move from an observed smudge to Lord Rosse's clawed nebula requires considerable magnification. A Lumicon Oxygen III (O-III) filter may bring filaments into view, but it remains an elusive target.

Images such as that shown here, though, gives a clear sense of the power driving the Crab. At its heart lies a pulsar – we will look at these objects in a moment – and the pulsar drives a process that produces visible filaments. In images taken by space-based telescopes, the Crab is a wonderful object.

And it has unexpected other uses. On January 5, 2003, the planet Saturn and its encircling moons passed in front of the X-ray bright Crab Nebula. The last time this happened was probably January 1296, when the Crab was presumably too small to be occulted. Also, the Chandra X-Ray Observatory was not available to watch it.

Chandra is a space telescope in orbit some 86,500 miles above Earth, designed to detect X-ray emissions from hot regions of the universe. Although Saturn is as much as a million times brighter than the Crab, the X-rays pounding from the heart of the nebula make it far and away the stronger source at those wavelengths. Observing the X-rays through the Saturn system made it possible for Chandra scientists to achieve a unique measurement of the thickness of the atmosphere surrounding Saturn's largest satellite, Titan. This was a key piece of information. Titan is the only satellite in the Solar System with an atmosphere, and it could not be acquired during the Voyager flybys of the moon. Chandra was able to establish that Titan's atmosphere was some 880 km thick, with a margin of error of about 60 km to either side.

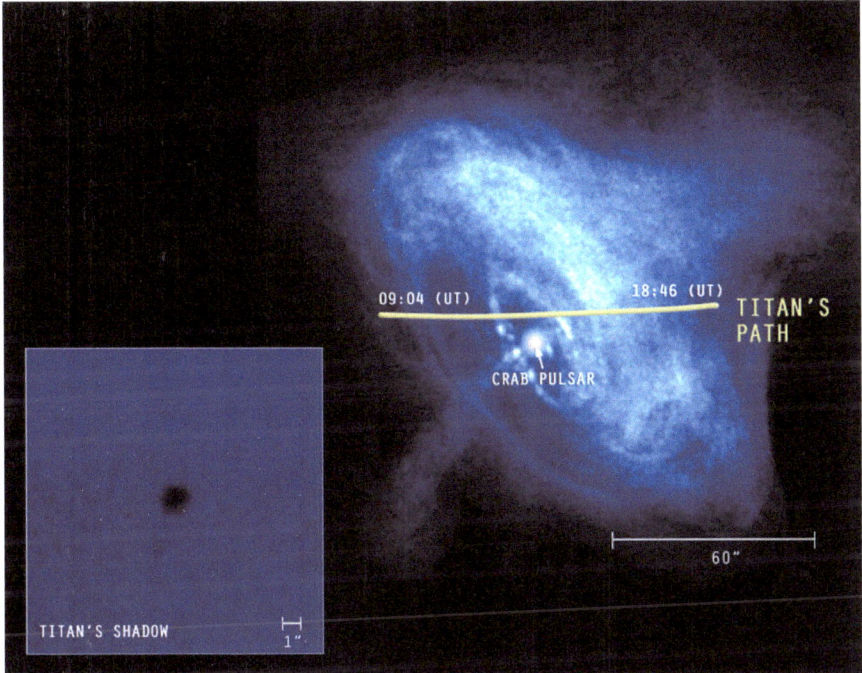

Fig. 17.3 Titan occults the Crab (Credit: NASA/CXC/Penn State/K.Mori et al.)

Lifting the Veil

A second supernova remnant observation lies some 1,500 light years away in the constellation Cygnus with its distinctive cross shape and the star Deneb at its peak. To find the Veil Nebula, composed of three sections labeled NGC 6960, 6992 and 6995, start with the arm of the cross pointing towards Pegasus and the star Epsilon Cygni. A little beyond bright Epsilon but along the same arm is Zeta Cygni. The Veil Nebula is just below and to the southwest of a line between these stars. Alternatively, if you have a good sense of the stars here, head straight for 52 Cygni and the easiest segment of the Veil to observe, NGC6960.

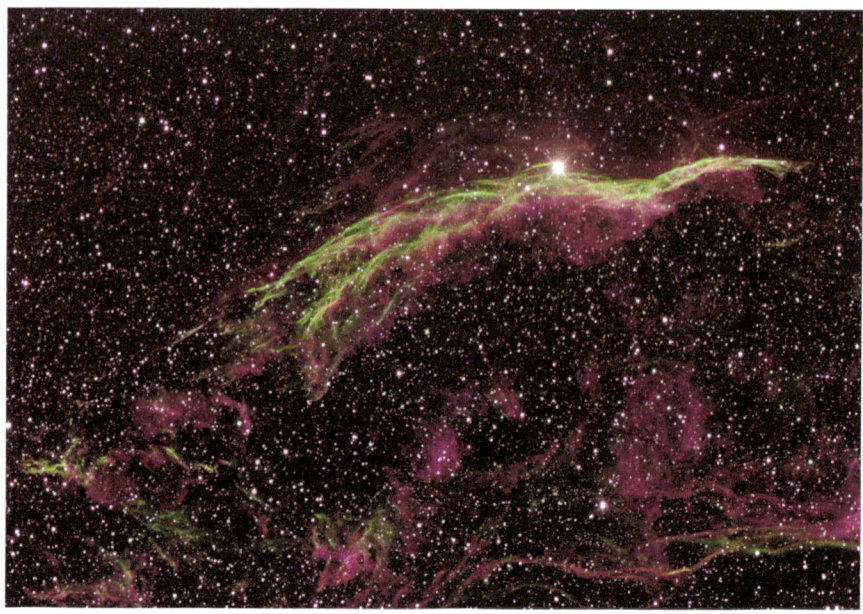

Fig. 17.4 The Veil Nebula, West (NGC 6960), captured on 7 August 2006 under a 95 % full moon with a 4" refractor (Credit: Keith Schlottman, http://www.xanaduobservatory.com)

With a telescope, no intervening moonlight and an OIII filter, this deep sky object is a wonderful target and, like the Crab, a supernova remnant. We are looking into starry death. In the case of the Veil, the original star exploded some 5,000–9,000 years ago.

Little Green Men

What we are seeing when we observe the Veil or the Crab are the remains of the explosion, outer layers of the star forced into the interstellar medium. This can happen only with stars larger than about 11 times the mass of the Sun, where conditions in the core allow for the formation of iron group elements. A supernova is a furnace for the creation of heavy elements – this is the r-process, where r stands for rapid – and while it is not the only way such elements are formed (there is also an s, or slow process, in lower mass giant stars), it seems that supernovae account for almost all of elements such as gold and plutonium.

The core of the exploded star remains. It is dominated by neutrons that exist in densities hard to imagine. These neutron stars are among the strangest objects in space. But stranger still were the pulsations detected in 1967 by Dame Jocelyn Bell Burnell and Antony Hewish, using a radio telescope near Cambridge in the UK. Searching for quasars, they identified unfamiliar signals so regular that, according to Jocelyn Burnell: "It was suspiciously like a man-made signal, but when we found that it kept a fixed place among the stars that seemed to rule that out. We dubbed it LGM, for Little Green Men."

Were these regular pulses of radio energy a signal from a distant culture? It turned out that Burnell and Hewish had discovered a new type of star. Pulsars such as that at the heart of the Crab Nebula function almost like lighthouses, swinging an immense magnetic field to create a regular beam of radio energy. Well over a thousand of these pulsars are now known in the Milky Way, all believed to have been formed out of Type II supernovae.

Some neutron stars do not develop in this way, and these are much harder to detect. Comprising a mass a few times the size of the Sun, these extraordinary bodies are packed into spheres as small as 10 km across.

The first human sight of the Crab Nebula can be dated with extraordinary accuracy thanks to observations preserved in a Chinese treatise on astronomy, the Sung-Shih, and on cave paintings made by native North Americans. This is the Chinese record of the seeing of a star, "as visible as Venus": "In the first year of the period Chih-ho, the fifth moon, the day chi-ch'ou, a guest star appeared approximately several inches southeast of T'ien-kuan." It dates the appearance of the Crab to July 4, 1054, and the supernova explosion that caused it to around 5446 BCE.

If it was as bright as Venus, was it seen in the Arab or European worlds? Scarcely at all. A chronicle reports a "circle of extraordinary brilliance" that appeared earlier in the year 1054 as a portent of the death of Pope Leo IX. Another refers to a "very clear star" appearing 4 years later in 1058. These might be misplaced references to the supernova. Meanwhile, cave paintings in the American Southwest depict images that might be a visual record of the event.

A similar pattern emerges in records of the supernova of 1181. There are three Chinese and five Japanese records of the star; "like Saturn and in color it was bluish-red and it had rays," according to one Japanese source.

Astronomers all over the world did observe the supernova of 1006. At least nine separate records survive in Chinese sources, which identify it as a prophet of prosperity for the state. "It was like Mars," a Japanese chronicler recalls, "bright and scintillating." It was also observed from the Arab world, "brilliant like the light of the Moon," and it was seen from the monastery of St. Gall in Switzerland, "glittering in appearance and dazzling the eyes." The relics of this supernova have been cautiously identified as a radio source in the constellation Lupus.

The 1572 supernova was also widely observed. Appearing in Cassiopeia, it far outshone the stars of its w-shaped asterism. "As large as a cup," one Chinese source records, "and its light rays were in all directions." Tycho and Francesco Maurolyco agreed, as we have seen. Queen Elizabeth I of England sought advice from her astrologer, Thomas Allen, to explain what the star might mean.

This debate over meaning reached a decisive moment some 30 years later, with another supernova. Now known as SN1604, this new star appeared in the constellation Ophiuchus the Serpent-Bearer in October 1604. It was observed on October 2, 1604, by Johannes Kepler and by Helisaeus Roeslin (1544–1616) and by Galileo from Padua some 8 days later.

As has been seen in earlier chapters, Galileo embraced the Copernican system as early as 1597, 13 years before he turned his telescope upon the stars. But he could see the 1604 supernova without need of any optical aid, and he incorporated his ideas about what a 'new star' could mean in his public lectures at Padua that year.

Roeslin was an astronomer but a committed anti-Copernican. Yet even he was not convinced of the complete changelessness of the sky. He observed the new star close to a conjunction of Mars with Jupiter and Saturn, a planetary alignment heavy with astrological import. But Roeslin was distracted by: "A new star, almost as large as Jupiter, but brighter, and as it sparkles it scintillates, like a fixed star, although in that place there is none."

What could such a new star mean? Roeslin associated the nova with politics, an astral message warning against the unification of the Scottish and English crowns and the independence of the Netherlands. If that was not enough, he believed it was also a sign of the end of the world.

Still, Roeslin recognized that the 1604 supernova changed the shape of the sky. It was the end, in his words, "of slavish submission to the doctrine of a single person." He meant by this that a new star could not be contained within the theories of the heavens that, coming from Aristotle and others, had dominated European and Arabic science for more than 1,500 years.

Why did that matter? Because it gave conceptual space for Copernicanism to grow. If one model had clearly failed, there was room for something new to take its place.

Kepler saw the new star as well. He made a rather different point. Writing in 1604, he said that: "The star's significance is a difficult matter to establish, and we can be sure of only one thing: that either the star signifies nothing at all for Mankind or it signifies something of such exalted importance that it is beyond the grasp and understanding of any man."

Nothing at all or too much to understand. If Roeslin's reluctant recognition of a new star opened the skies to a new way of understanding them, then Kepler used the same event to reclaim the same skies for science from astrology and the instinct to interpret events in the heavens as about human beings on Earth. Cosmological events could be observed and analyzed – their meaning, if any, was unknown. This was, as he put it, "no ordinary thing, like throwing dice."

Kepler's words did not end the practice of astrology, nor was that his intention. Kepler himself maintained an astrological practice because it gained him money and patrons. Astrology was part of the toolbox of early modern astronomers. Kepler did, though, admit in 1616 that it was "only a little more honest than begging." Yet it was to be a significant long-term consequence of the Copernican revolution that the stars acquired lives and deaths that had nothing to do with human fears.

Fig. 17.5 Chandra image of the Tycho supernova remnant (Credits: X-ray: NASA/CXC/ Rutgers/K. Eriksen et al.; Optical: DSS)

The supernovae of 1572 and 1604 can still be seen. Their remnants continue to expand. The image shown, from the Chandra Observatory, reveals low-energy X-rays in red and high energy X-rays in blue continuing to expand in a blast wave from the 1572 supernova observed by Tycho Brahe. These reveal stripes never previously seen in a supernova remnant. An observation first made over 400 years ago continues to excite questions among scientists. We are no longer asking for meaning but wanting to delve deeper into how they work.

Chapter 18

Habituated to the Vast

Observation: Uranus and Neptune
Significance: The expansion of the Solar System, the Voyager mission
Science: Ice giants, their atmospheres and satellites

For all but a sliver of human history, the boundaries of the Solar System have been set by our eyes. Saturn, the most distant naked-eye planet, was where the power of the Sun was deemed to end. Beyond it, the ancient and medieval models of the cosmos showed the fixed sphere of the stars and then the outer limit of all things.

This chapter is about the breaking open of that vision of the heavens. The discoveries of first Uranus in 1781 and then Neptune some 50 years later were together a further revolution in our understanding of space. Like that effected by Galileo's observations, it was achieved by new and bigger telescopes. After the brief human history of these planets, we will turn to observations and modern science, before concluding with a look at the only human-built machine ever to reach these far-distant worlds.

Many a Dish of Coffee

When you do observe the planet, you will understand why people probably saw Uranus before they recognized it. Galileo himself might have done so. Should we say of them, "They got it wrong," when in a few clicks we can decorate our desktop with its image? Or should we rather recognize that a lot of science consists in getting things wrong, in observing with the tools available and making reasonable conclusions while amassing more evidence? The latter must be right.

William Herschel started by getting Uranus wrong. When on March 13, 1781, he used his 7-ft telescope with a 6-in.-diameter and reflecting mirror that he built

M. Marett-Crosby, *Twenty-Five Astronomical Observations That Changed the World: And How To Make Them Yourself*, The Patrick Moore Practical Astronomy Series, DOI 10.1007/978-1-4614-6800-4_18, © Springer Science+Business Media New York 2013

himself to spot a small star between Auriga and Gemini, he supposed it was a comet. *An Account of a Comet* was how he titled the paper he presented in April, a careful piece of work full of evidence laid out in tables. "Many a dish of coffee during the long nights of watching," his sister Caroline recalled that they had needed as the observations carried on.

On April 6, the comet misbehaved. The Herschels recorded that, "With a magnifying power of 278 times, the Comet appeared perfectly sharp around the edges, and extremely well defined, without the least appearance of any beard or tail." A little later, William wrote to Astronomer Royal Nevil Maskelyne that he was observing: "A comet or a planet that is very different from any comet…. I ever saw."

Or a planet – Herschel was careful; he knew what his words might mean.

In later years, Herschel faced the charge that he had made his revolutionary discovery by accident. In one sense, he had. Herschel had not been looking specifically for a new planet. But writing to Sir Joseph Banks in November 1781 Herschel explained how good astronomy works:

> *The new star could not have been found out even with the best telescopes had I not undertaken to examine every star in the heavens including such as are telescopic, to the amount of at least 8 or 10 thousand….The first moment I directed my telescope to the new star, I saw with a power of 227 that it differed sufficiently from other celestial bodies; and when I put on the higher powers of 360 and 932 was quite convinced it was not a fixed star.*

In other words, he looked until he understood.

He also thought that he saw red rings around the planet. Later observations ruled this out. His "newly polished speculum, of an excellent figure," which Herschel used on March 5, 1792, showed "the planet very well defined, and without any suspicion of a ring." In fact, Uranus does have rings, but there is no likelihood that Herschel could have identified them. He took a calm view of the matter, stating: "The observations which tend to ascertain the existence of rings not appearing satisfactorily supported, it will be proper that surmises of them should be given up, as ill founded, or at least reserved till superior instruments provided, to throw more light upon the subject." Superior instruments eventually did just that, but it was to take nearly two centuries.

What Herschel did observe was that Uranus possessed satellites. He thought he saw six, but four of these proved to be stars. Nevertheless, the Uranian moons later named Titania and Oberon are genuine Herschel discoveries. Ariel and Umbriel were discovered in 1851 by the English astronomer William Lassell (1799–1880).

Lassell was a dedicated planetary observer who built his own 9-in. equatorial reflector with a mount specifically designed for the precise movements needed to track Solar System objects. He later constructed a 24-in. device, which he took to Malta, an early example of choosing a suitable dark sky location for a big telescope.

Herschel's discovery more than doubled the size of the Solar System. Uranus orbits at an average distance of 1.78 billion miles from the Sun, about 19 times farther from the Sun than is Earth. It takes 84 of our years to complete an orbit that, for some time after its discovery, seemed to defy Newton's laws of gravity and motion.

Herschel was sober about what he had achieved. He followed Galileo's example in offering the planet to his monarch, proposing that the planet be named Georgium Sidus, George's Star, after British King George the Third. Those less indebted to the monarch of Windsor Castle urged a supranational identity, and the planet eventually acquired the patronage of the Greek muse of astronomy, Urania. Herschel, equipped with a royal pension and larger telescopes, devoted himself thereafter to the study of deep sky objects and cosmology.

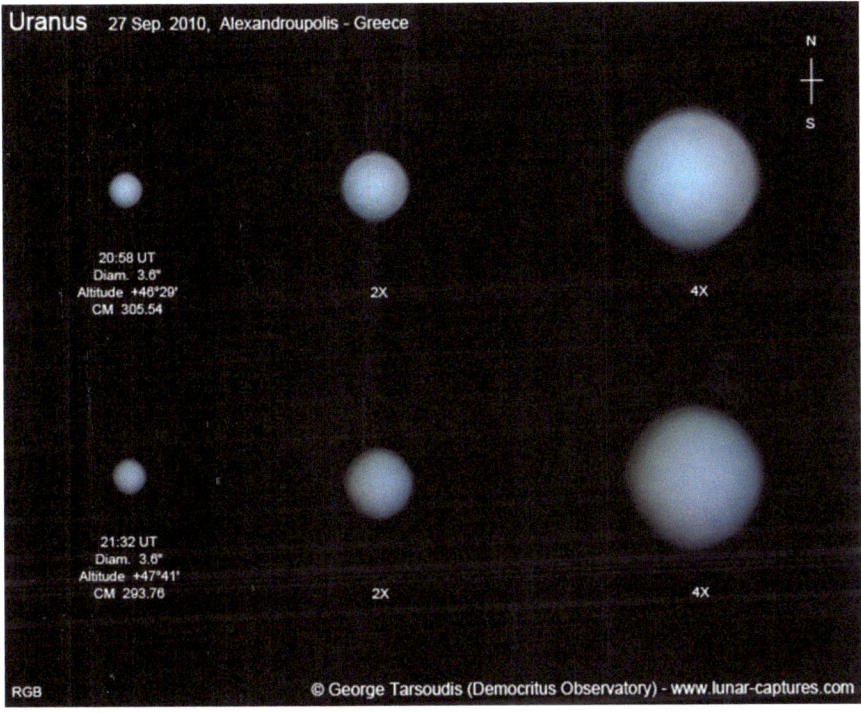

Fig. 18.1 Images of Uranus captured on 27 September 2010, magnification increased. Telescope Newtonian Orion Optics 250 mm @f/6.3, Unibrain camera Fire-i 785, barlow 3X, Filter Red, Green, Blue (Credit: George Tarsoudis, www.lunar-captures.com)

Observing Uranus

Many astronomical magazines and websites show you where to start looking for Uranus. As with all planetary observing, it is worth identifying when Uranus is at opposition, rising in the east at sunset and remaining in the sky all night. Uranus' magnitude varies much less than that of Mars and the gas giants, but with this planet, even a little can help.

The best way to search for Uranus is with binoculars. In the lifetime of this book, Uranus will be passing among the southern stars of Pisces, drifting in the direction of Aquarius. This places it beneath the Square of Pegasus, the four corner stars of which are much brighter than the planet. On the eastern side are the two stars Algenib (Gamma Pegasi) and above it Alpheratz. A line between these two stars is an approximate, but only approximate, pointer to where the planet can be found over the next few years. Most of the stars in Uranus' apparent neighborhood on the edge of Pisces are fainter than the planet's +5.7 magnitude.

Once you have found Uranus, slowly increase the magnification. The planet will appear much like a star in small instruments, but larger telescopes will reveal its characteristic 'fuzzy tennis ball' appearance. In good conditions, it might even glimmer green.

It is technically the case that four of Uranus' 27 known satellites can be seen with an amateur telescope. It is possible with an 8-in. aperture to spot, just, Titania and Oberon, but Umbriel and Ariel are very close to the planet's glare. The only way to find faint targets like these is to know beforehand where to look by consulting a satellite diagram. These are available in online and printed sources. A really tricky observation in imitation of those we made with the Galilean moons is to seek out a transit event within the Uranian system, when tiny shadows from its moons cross the planet's disk. These can be predicted via online tools.

However, the main joy of observing Uranus is finding that it is there. Herschel's telescope uncovered a Solar System beyond the reach of human sight. Our telescopes can take us to that same place.

What Is Uranus Like?

When the *Voyager 2* spacecraft passed by the planet in 1986, the close-up view was bland. But don't let this is deceive you. The science of the planet is both alluring and strange.

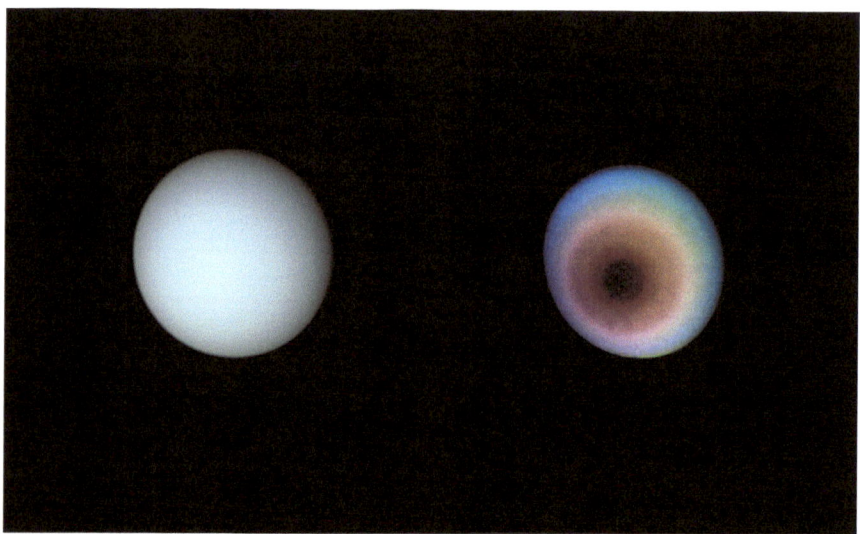

Fig. 18.2 True-color (*left*) and false-color views of Uranus, taken by Voyager 2 on January 17, 1986, range 5.7 million miles (Credit: NASA)

Uranus and Neptune are giant planets, both like and unlike the gas giants Jupiter and Saturn. Specifically, they are ice giants, composed of a layered interior that is probably simpler than that of their larger cousins. They are large bodies. Uranus is more than 14 times the mass of Earth and Neptune 17 times, but this becomes paltry beside Jupiter, which weighs in at 318 times our size.

By contrast, Uranus is almost equally dense and Neptune actually denser than Jupiter. This is an important measurement, telling us a great deal about their internal structures. Theory suggests that they started life as kernels of rock, but there is no direct evidence that the planets have differentiated cores now. Rather, their densities suggest that icy materials and rock become more compressed towards their center, but have not completely separated. There is also a higher admixture of other elements with the hydrogen and helium. These gases predominate in the outer layers of the ice giants, with icy materials below them.

Note that the words we use to describe giant planet interiors carry more meaning than we often intend. Neither *icy* nor *rocky* bear the implications that they do on Earth. Gas giants are not all gas and the ice is not ice as it might come out of a freezer. The state – solid, liquid or gas – of any particular constituent in the internal mixture is hard to assess. We have no way of taking samples and, for Uranus as for the other giants, there is no meaningful 'surface.' Even our word 'layer' can mislead. The boundaries within the planet are probably not delineated sharply.

Both ice giants have near identical internal temperatures. This is comforting at first glance, suggesting that they work in the same way, until we recall that Neptune is far further from the Sun. This means that, while Neptune emits more heat than it absorbs, like the other giants, Uranus does not. Yet both have magnetic fields, implying hot interiors.

There are two possible explanations for Uranus' chill – both answers pose new questions. The planet might have some interior structure that restricts the emission of internal heat. Alternatively, the planet started off by being colder due to some unique aspect of its formation and has been losing heat at the normal rate but from a lower base.

A further aspect of the planet that we cannot properly explain is its magnetic field. This is tilted from the rotational axis such that auroras occur away from the rotational poles, in contrast to the situation on Earth and other planets. Observations made by the Hubble Space Telescope in 2012 compounded this mystery by imaging auroral events not as curtains of light but as spots or dots.

What causes Uranus' magnetic field? It's existence implies either an iron-rich core, which seems unlikely if our model for the planet's interior is right, or a large liquid layer. As with so much of the science of this planet, we have to conclude that, thus far, we do not know.

The cold and fuzzy tennis ball Uranus is also lying on its side, tipped by a massive 97.9 °. The effects of this include an equator running from top to bottom and poles that can face the Sun. When *Voyager 2* passed by the planet, it found more solar energy at the pole than the equator. A season on Uranus lasts about 21 years, and some regions receive no sunlight over very long periods. This has consequences for planetary weather, cloud cover and wind speed that are not fully understood.

This tilt, unique in our Solar System, is now stable. But what caused it? One scenario envisages a collision between a young Uranus and an Earth-sized proto-planet. It's a dramatic account, but does not explain why, in contrast to the planet's oddness, its regular satellites orbit contentedly around the equatorial plane. An alternative model posits an interaction between Uranus and a large migrating planet that dragged Uranus out of kilter in some way.

The atmosphere is predominantly hydrogen and helium, with smaller amounts of methane and clouds caused by hydrocarbons reacting with sunlight. Though it is spread across the planet in such a way as to appear uniform, there is evidence for both atmospheric banding and clouds.

Fig. 18.3 Dark clouds in the atmosphere of Uranus, imaged by the Hubble space telescope (Credit: NASA, ESA, L. Sromovsky and P. Fry (University of Wisconsin), H. Hammel (Space Science Institute), and K. Rages (SETI Institute))

When the Voyager probe passed Uranus in 1986, there was no visible activity within the clouds. At that time Uranus was a settled, uniform world. Now it seems that the planet was merely slumbering. Observations made with the Hubble Space Telescope and at the Keck Observatory in Hawaii have identified dark clouds as in the above Hubble image and a distinctive dark spot and accompanying bright companion area. The dark spot, analogous probably to that seen by Voyager on Neptune, may be a long-term or an unstable feature, but it is clearly a storm spinning through the atmosphere. It has not yet been possible to determine its detail – it might be a dark cloud, a hole in a very bright cloud or the topmost marker of much deeper atmospheric turbulence. But during the 2006 observing period it showed notable variations in contrast.

What does seem clear is that the slow progress of Uranus' seasons does create changes. As its long winter ends, so one new storm at least is puncturing the once quiescent atmosphere.

Uranus is however more like its neighbors in that rings encircle the planet. At least 13 distinct rings have been identified, many of these associated with shepherd moons. There is no possibility of observing these rings with the equipment at our disposal, but they are occasionally visible to professional telescopes when end-on to Earth.

The Satellites of Uranus

There are 27 known satellites of Uranus, five in regular orbits around the planet. The others, many identified by *Voyager 2,* have eccentric orbits and are probably bodies captured from space by the planet's gravity.

Titania is the most massive of the satellites; Voyager images show a complex and bombarded world. Oberon is similarly cratered, but dark material from the interior has leaked onto the surface. Umbriel is also dark, its water-ice signal is weaker than for any of the other satellites. The Voyager probe came closest to Ariel, which shows some bright areas on its surface and signs of recent geological activity, presumably caused by mechanisms of tidal heating.

Miranda is the smallest and innermost of Uranus' moons. No one expected much of it before the Voyager flyby, but as the probe approached, images revealed a complex and evolved surface, cratered in part but with distinct regions of grooves, labeled the Arden, Elsinore and Inverness Coronae. There are suggestions that this moon was previously in a different orbit from which it somehow escaped, and that the stresses caused by this mobile past have resulted in greater internal heating. An alternative theory posits a massive impact shattering the moon, which then re-formed into the complex body that exists today.

It has been suggested recently that the relationship between Uranus and its satellites has been chaotic and that the rings were perhaps created by continual collisions between small moons. *Voyager 2* images are being re-examined. In one, evidence has been found of a moon not noticed in earlier studies.

There are no probes scheduled to visit either Uranus or its moons. Observation is difficult. We will continue to wonder at what Uranus really is like.

Finding Neptune

Observers of Uranus were surprised at the shape described by the planet's orbit. It appeared to slow down, something Newtonian laws of motion could not permit. The French astronomer Alexis Bouvard (1767–1843) stated: "leave to the future the task of discovering whether the difficulty...results from the inaccuracy of the ancient observations, or whether it depends on some extraneous and unknown influence which may have acted on the planet."

The challenge of predicting this *extraneous and unknown influence* was taken up in the 1840s by Urbain Le Verrier (1811–1877) and John Couch Adams (1819–1892). Adams, aged only 22, determined to solve the problem of Uranus' odd orbit, making this note to remind himself of his intention:

1841. July 3. Formed a design, in the beginning of this week, of investigating, as soon as possible after taking my degree, the irregularities in the motion of Uranus...in order to find whether they may be attributed to the action of an undiscovered planet beyond it; and if possible thence to determine the elements of its orbit...which would probably lead to its discovery.

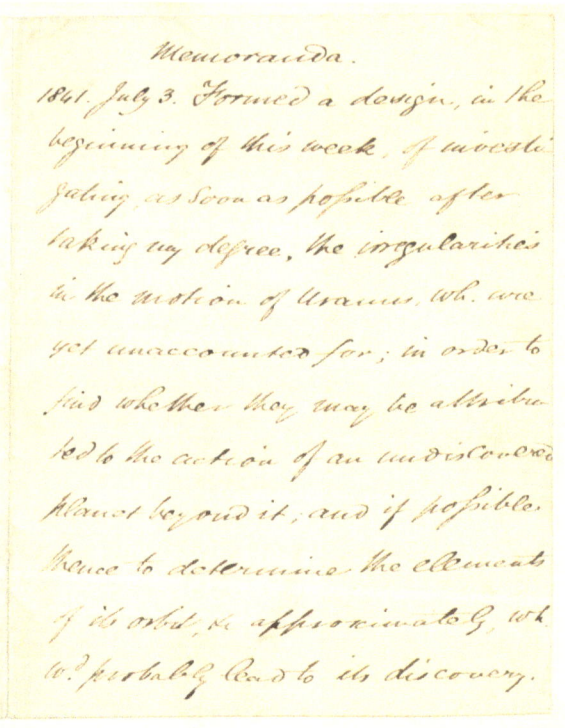

Fig. 18.4 The John Couch Adams memorandum (Credit: By permission of the Master and Fellows of St John's College, Cambridge)

Neptune was the fruit in the first instance of mathematics more than the telescope. Adams and Le Verrier plotted where the mystery planet should be. It was Le Verrier's work that met with observational success when he dispatched his calculations to the Berlin Observatory. On September 23, 1846, Johann Gottfried Galle (1812–1910) and Heinrich Louis d'Arrest (1822–1875) used the 9½-in. Fraunhofer refractor to identify an eighth magnitude star not far from Le Verrier's predicted location. "That star is not on the map!" d'Arrest is said to have called out. Two days later, Galle wrote to Le Verrier: "The planet of which you indicated the position really exists.... We are thereby, thanks to you, definitely in possession of a new world."

Neptune appears to the amateur observer much as it did in 1846 – an inconspicuous, eighth magnitude star. It requires considerable magnification to persuade it out of its anonymity such that a tiny blue disk can be made out. Magazines and online resources indicate where Neptune is to be found, and observing it reconnects us to the magic of this moment. Many threads of the scientific revolutions of previous centuries – the Galilean telescope, Keplerian orbits, Newtonian mathematics – found a communion on Neptune's distant shore.

Neptune's largest moon is Triton, 2,700 km in diameter. It is the only substantial satellite of any planet that orbits in the retrograde direction. This has prompted some scientists to suggest that Triton originally orbited the Sun. It might be a captured planet imprisoned, as it were, into satellite status. Alternatively, it may have been grabbed by Neptune out of a binary pair of Kuiper Belt Objects.

Bringing It Back Together

A few small numbers: 722 and 815 kg, 68 K of computer memory, two cubes 4 m on each side. And then two large ones: 18,009,358,894 km (120 AU) and 14,676,539,775 (98 AU). These last are the distances traveled from Earth at time of writing by *Voyager 1* and *Voyager 2*. The earlier numbers are their weights, processing power, and the size of the boxes you would need to pack them up. By any measure, the grand tour of the two Voyager spacecraft is one of the great achievements of human science.

Their mission was ambitious from the start. The project plan intended that: "The spacecraft will continue to escape from the solar system toward the solar apex, the communication could be maintained as long as the spacecraft continues to function. If the spacecraft continues to function past Saturn encounter, an extended mission could be conducted in anticipation of penetrating the boundary between the solar wind and the interstellar medium."

That is what they are still in the process of achieving. Launched in 1977, the two probes have explored Jupiter, Saturn, Uranus and Neptune before passing into the heliopause. Bits of the spacecraft have broken, but they've been tended from Earth with such love that five instruments are still working on *Voyager 1*, six on *Voyager 2*. Their 3.66 m diameter antennae have communicated images of huge scientific significance and remarkable beauty.

The Voyagers have done in the late twentieth and twenty-first centuries what Herschel, the mathematicians who predicted Neptune and the astronomers who discovered it did in their own time. They have measured the extent of the solar family.

The Voyagers are also a message. Each probe has an LP record – remember those? – stuck to its front, bearing 118 images from Earth, greetings in 54 languages, music and other sounds. Instructions on how to use the records are etched into the housing. This is who we are and where we live, they say, summarized in analog. Let's hope that this technology, lost to most of us, is remembered wherever the Voyagers end up.

They also send a message back.

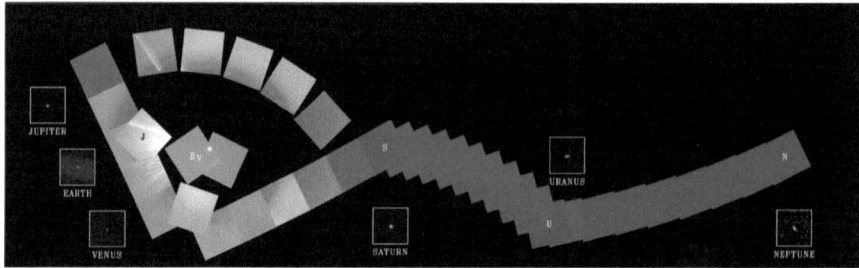

Fig. 18.5 The cameras of Voyager 1 on Feb. 14, 1990, pointed back toward the sun and took a series of pictures of the sun and the planets. In the course of taking this mosaic consisting of a total of 60 frames, Voyager 1 made several images of the inner solar system from a distance of approximately four billion miles and about 32 ° above the ecliptic plane. Thirty-nine wide angle frames link together six of the planets of our solar system (Credit: NASA/JPL)

The Sun is the bright object marked S while Earth is, in Carl Sagan's famous phrase, "a pale blue dot." Sagan went on to write:

> *From this distant vantage point, Earth might not seem of any particular interest. But for us, it's different. Look again at that dot. That's here, that's home, that's us. On it everyone you love, everyone you know, everyone you ever heard of, every human being who ever was, lived out their lives. The aggregate of our joy and suffering...every superstar, every supreme leader, every saint and sinner in the history of our species lived there – on a mote of dust suspended in a sunbeam.*

Fear and Wonder

This image dwarfs us. Earth becomes depopulated and also de-nationed, geographical much more than human. It also confuses – astronomy is meant to bring the unexplored parts close and not to squash us into less than a single pixel.

This is how Samuel Taylor Coleridge (1772–1834) remembers seeing the skies around the time of Herschel's discovery of Venus:

> *I remember that at 8 years old I walked with him [Coleridge's father] one evening...and he told me the names of the stars and how Jupiter was a thousand times larger than our world and that the other twinkling stars were suns that had worlds rolling round them....I heard them with profound delight and admiration, but without the least mixture of wonder or incredulity. From my early reading of fairy tales...my mind had been habituated to the vast.*

Habituated to the vast. Our observations of Neptune and Uranus are, like all astronomical undertakings, about both answers and new questions. The same is true of the Voyager family portrait. Is our Solar System a rare or common gathering? is life unique, occasional or common? The answers we give to these questions in our own time will in all likelihood turn out to be wrong. You will not have got this far in these 25 observations without discovering how astronomy often involves making mistakes.

We end with William Herschel. He argued for some time, and in good company, that he could see evidence of life on our own Moon. He changed his mind as he observed more frequently and carefully. But he wrote to Nevil Maskelyne: "My saying that there is an absolute certainty of the Moon's being inhabited may perhaps be ascribed to a certain Enthusiasm which an observer, but young in the science of astronomy, can hardly divest himself of when he sees such wonders before him."

Enthusiasm in the face of wonder. The fuzzy tennis ball inspires us to look further. There must be more to see.

Chapter 19

Monsters
of Magnitude and
Invisible Nothings

Observation: Cygnus X1, Canes Venatici, M51
Significance: The extreme fate of matter, the way galaxies work, the violence of space
Science: Black holes

"The universe has taken on a wholly new face. It has become more exciting, more mysterious, more violent and more extreme."

So wrote Isaac Asimov in 1980, introducing to readers of the *Saturday Evening Post* something with "the simplest, plainest, calmest and mildest name – nothing more than a black hole."

"Nothing more" is about right. Black holes are among the strangest objects in the sky. They tear apart that sense of peacefulness that the stars seem to give us, a study of the violence of the universe. Observing and understanding them gives insight into the way that galaxies work, taking us towards "the most extreme state of matter possible," as Asimov perceived.

They are true to their name, absolutely unobservable in optical wavelengths. This is the only chapter where failure to see the observing target means that you have been successful. The best that optical telescopes can offer is proximity.

Black Swan

Getting near a black hole starts at the northern cross asterism in the constellation Cygnus, a cruciform of five stars with Deneb at the upper end near Cepheus and Albireo at the lower end close to Vega and the constellation Lyra. Deneb marks the tail of the celestial swan, and were it not 3,000 light years away, the star would

M. Marett-Crosby, *Twenty-Five Astronomical Observations That Changed the World:* 229
And How To Make Them Yourself, The Patrick Moore Practical Astronomy Series,
DOI 10.1007/978-1-4614-6800-4_19, © Springer Science+Business Media New York 2013

dominate Earth's hours of darkness, for it is one of the most powerful stars in the galaxy, a white A-type supergiant, a rare species in the sky, whose apparent magnitude from Earth of +1.25 only hints at its vast size. Its absolute magnitude is −7.1, and it has a radius of 110 suns, extending the star, were it in our Solar System, half way towards Earth. Albireo is a visual binary, straightforward to split, comprising a yellow K-type giant and B-type pale blue companion. It is perhaps the most beautiful double star in the sky.

The core of Albireo's name comes from Ptolemy, who labeled this constellation Ornis, the Bird. Confused Latin translators thought that this name came "from that of a herb," *ab ireo*. These two words were joined together, and an *l* added to make it sound more Arabic, a muddle that created our modern Albireo.

Fig. 19.1 Splitting Albireo (Credit: Frederick Ringwald)

The evidence of the Milky Way is all over Cygnus. Between Deneb and the arm of the northern cross that stretches towards Pegasus lies a dark patch known as the Coalsack, which at its northeast corner becomes NGC7000, the North American Nebula. Close alongside Deneb, this beautiful deep-sky binocular object is a gas cloud glowing with the fire of young suns. Moving to the star at the center of the cross, Gamma Cygni, or Sadr, a night of good seeing should reveal the Cygnus Star Cloud, a span of light dividing the Coalsack from the Great Rift, the dark streak stretching from Albireo. This darkness is formed from dust clouds within the Milky Way system.

The line from Gamma Cygni down to Albireo, forming the body of the cross, passes our first black hole. If you follow the asterism in binoculars, you will spot about half way between the two bright stars a third star, Eta Cygni. There is nothing too remarkable about this star except that it lies close to Cygnus X-1, the first candidate black hole for which there is strong evidence.

The discovery of Cygnus X-1 can be credited to a remarkable satellite, the X-Ray Explorer Satellite SAS-1 known by the Swahili name Uhuru, meaning freedom. Launched on December 12, 1970 from the San Marco platform off Kenya on that country's Independence Day (hence its name), Uhuru's two and a half year mission identified 339 objects emitting powerful X-rays. Principal among these was the object later identified as our Cygnus X-1. Its strong emissions argued for a compact object, and it was too intense a signal to come from a white dwarf, while microsecond fluctuations in the X-rays suggested that it was not a neutron star. It seemed to be an X-ray source about 10 times the mass of the Sun with a companion massive star.

Riccardo Giacconi, one of the scientists behind the Uhuru mission and awarded the Nobel Prize in 2002 for his contribution to X-ray astronomy, wrote: "The Cygnus X-1 X-ray emitter is a compact object of less than 30 km radius due to the rapidity of the pulsations and the fact that the pulsations are so large that they must involve the whole object. The object has mass greater than that allowed by our current theories for neutron stars. Therefore the object is the first candidate for a black hole."

Just a candidate? Yes. Despite the way that black holes have entered common culture, astronomers remain cautious about saying for certain that any one object is a black hole. What we can say of Cygnus X-1 is that it fits the theory. Let's now turn to what that theory tells us.

No Escape

Black holes have an unexpected scientific history. The term was first used by the American physicist John Archibald Wheeler (1911–2008) in 1968, but the concept is two centuries older. It was first proposed in 1784 by English clergyman John Mitchell (1724–93). Described as "a little, short man of black complexion, and fat," Mitchell was a student of Newton's theory of gravity. It was by reflecting on the implications of Newton's work that Mitchell came to the conclusion that an object from which light could not escape might exist. This is his argument, as presented in the *Philosophical Transactions of the Royal Society* for 1784:

> *If the semi-diameter of a sphere of the same density with the sun were to exceed that of the sun in the proportion of 500 to 1, a body falling from an infinite height towards it, would have acquired at its surface a greater velocity than that of light, and consequently, supposing light to be attracted by the same force....all light emitted from such a body would be made to return towards it, by its own proper gravity.*

All light emitted from such a body would be made to return towards it. Mitchell perceived that it was inherent in Newton's model of gravity that, in this most extreme case, gravity would conquer light. The Marquise de Laplace drew the same conclusions in 1796.

Both Mitchell and Laplace were exploring escape speed. They recognized that, as the mass of an object shrinks, so the velocity required for it to escape gravity must increase. When that velocity is required to be greater than the speed of light, there can be no release.

The equation for escape is elegant and not hard to understand. Written formally, it states $v_{esc} = \dfrac{\sqrt{2GM}}{R}$, where M is the mass of the body trying to escape, R, is its radius and G is the force of gravity. v_{esc} cannot exceed the value c, the speed of light.

So had Mitchell and Laplace created the modern theory of black holes? Both men were dealing with a theory of light as a stream of particles, and they assumed that these particles had mass. They don't. Their Newtonian tools, powerful as they were, could not express the true nature of black holes. That required Einstein's understanding of space and time.

In 1915, Karl Schwarzschild (1873–1916) worked out solutions to Einstein's field equations, establishing the radius of an object that, if its mass object were squeezed into that space, the escape speed would equal the speed of light. Known as the Schwarzschild radius, this defines the size of a black hole and states that they must exist in a universe of general relativity.

Where do they come from? The formation of a black hole is a consequence of the collapse of massive stars. If gravity overwhelms all forces opposing it, such as those which maintain a neutron star in being, the implosion continues until the star has been compressed into an object of vast density but occupying a tiny space, a singularity. Smooth and perfect, a black hole is bordered by a one-way membrane, the event horizon. An object falling beyond this line would undergo something called the noodle effect. Extended towards the center of gravitation, its sides would be crushed, turning whatever it once had been into something like a noodle. Body, cells and atoms would in turn be crushed. So, too, the protons and neutrons within the atoms. This is Isaac Asimov again: "For objects falling into a black hole there is no reaching the center, no zero volume, no infinite density – yet there is also no turning back."

Lying at the extreme of the Einsteinian universe, black holes have mass and spin and nothing more.

How can we find them? Primarily by detecting mass plunging towards them in the X-ray wavelengths and by looking where we think they might be found. Their most significant haunt is at the centers of galaxies.

Among the Hunting Dogs

If you look up at the asterism of the Big Dipper in Ursa Major, you will notice that the area of the sky contained within the arcing loop of the Dipper's tail is fairly empty, with just a single bright star before the outline of Boötes and its mighty star Arcturus. It was into this quiet patch of the night that Jonannes Hevelius

(1611–1687) inserted two hunting dogs, the constellation Canes Venatici. The brightest star is Alpha Canum Venaticorum, or Cor Caroli, named after British King Charles I. Alongside it in the direction of the Plough lies the Beta Canum Venaticorum, the star Chara. Cor Caroli is a binary star quite easy to separate, while Chara glimmers like a very distant Sun.

The star Alpha 2 (α^2) Cor Caroli has a very strong magnetic field and is the prototype of the Ap group of stars. Whereas the Sun's field strength is about 2 G, that of the star is 1,000 G.

From Chara, move northwards using binoculars towards the second last of the Big Dipper stars and you will encounter the star Y Canum Venaticorum, better known by the nickname granted it by Angelo Secchi as La Superba, one of the coolest and reddest stars in the sky. It seems to be leaking blood into the night.

La Superba is a variable star, easing between magnitudes of +4.8 and +6.3 over a period of 160 days. It is fusing helium into carbon and oxygen as it approaches the end of its life. But in La Superba, carbon production is outpacing oxygen, giving it that extraordinary redness that compares only with other carbon stars such as R Leporis and Mu Cephi.

La Superba is a distinctive guide towards a deep sky object where a black hole is thought to lurk. From the red star, move a little towards the falling bowl of the Big Dipper asterism, and you will come to galaxy M106. It is one of the closest Messier objects to Earth. A late addition to his catalog, it has a strong radio source at its core. Once again we think, but do not know, that this must be a black hole.

While in this constellation, it would be a shame not to observe this.

Fig. 19.2 M51, The Whirlpool Galaxy, imaged on 11 June 2007 with a C14 telescope and ST10-XME camera (Credit: Paul and Liz Downing, www.paulandliz.org)

The Whirlpool Galaxy M51 is a favorite target for astronomers, photographers and sketchers alike. Although in Canes Venatici, it is best found by returning to the Big Dipper and the star at the tip of the handle, Eta Ursae Majoris. If you drop down from this star towards Cor Caroli, you will pass across this face-on galaxy of some 200 billion suns, attached to its beautiful companion galaxy NGC5195.

Black Holes and Galaxies

While black holes span a range of masses, there seems to be good evidence that many galactic centers harbor massive or even supermassive black holes. Moreover, there is a correlation between the masses of these black holes and their galaxies. The black hole weighs 0.2 % of the mass of its host galaxy. Is this an accident or a false result? Or does it imply that massive black holes are the seeds of galaxies, and that the size of one affects the other? Either way, we are probing some of the deepest questions of cosmology here. Black holes have entered the human imagination and may provide a key to unlocking why the galaxy we live in is the way it is.

Our own galaxy, the bright streaks and dark clouds, which we observed earlier around the constellation Cygnus, has a powerful radio source at its heart. Known as Sgr A* or Sag A*, it seems to be one of the least massive supermassives, as it were. It is closer to us than any other galactic black hole, and so is a special place where we test our theoretical models of black holes.

Observing Sgr A* takes us back to Sagittarius, which we explored in an earlier chapter. It lies on the extreme edge of the constellation beyond the star Kaus Australis, where it borders Ophiuchus.

Remarkable observations of the vicinity of Sgr A* have revealed that bright stars surround the radio source. Many of them are young, between 2 and 8 million years old, one of the richest concentrations of massive young stars in the galaxy. They pose the question of whether they formed in the neighborhood of Sgr A* or were transported there. This 'paradox of youth,' as it has been termed, indicates how much we have to learn about stars in this extreme environment.

One of the signs of the Sgr A* black hole is the way that these close stars orbit the central point.

The fate of one such star is demonstrated in PS1-10jh, a galaxy some 2.7 billion light years from Earth. In mid-2010, astronomers using the Pan-STARRS1 telescope in Hawaii and the space-based Galaxy Evolution Explorer identified the turmoil caused by a star that had wandered too close to the suspected central black hole. Its orbit was very tight, and at its closest the star came within 6 times the Schwarzschild radius. The gravitational force of the black hole seems to have torn into the star. Part of the star was expelled at high speeds while within 2 months other remains had fallen beyond the event horizon. "It is like we are gathering evidence from a crime scene," lead researcher Suvi Gezari has said. "We detect from the carnage that the slaughtered star had to be have been the helium-rich core of a stripped star."

Some of the mysteries of the physics of black holes might be illuminated, quite literally, during 2013, when a cloud of dust observed by the VLT array in 2011 is expected to enter the accretion zone of Sgr A*. The black hole's event horizon seems to be going through a peaceful phase at present, but that is going to change as the low temperature cloud some five times brighter than the Sun continues to feel the black hole's power. The cloud is hurrying up – it is estimated to have doubled its speed over the last 7 years – and is being stretched in a manner not unlike that shown during the encounter with a black hole in *2001 – A Space Odyssey*. This is how Stefan Gillessen of the Max Planck Institute for Extraterrestrial Physics in Germany described it to the BBC: "The idea of an astronaut close to a black hole being stretched out to resemble spaghetti is familiar from science fiction. But we can now see this happening for real to the newly discovered cloud. It is not going to survive the experience."

Galactic Center Survey
Hubble Space Telescope NICMOS • *Spitzer Space Telescope* IRAC

NASA, ESA, Q.D. Wang (University of Massachusetts, Amherst), and S. Stolovy (Caltech)   STScI-PRC09-02a

Fig. 19.3 Hubble-Spitzer color mosaic of the galactic center (Credit for Hubble image: NASA, ESA, and Q.D. Wang (University of Massachusetts, Amherst). Credit for Spitzer image: NASA, Jet Propulsion Laboratory, and S. Stolovy (Spitzer Science Center/Caltech))

In These Our Sight Plunges

Black holes pose many questions – about themselves, the galaxies around them, about how matter became formed into galaxies at all. But what about those who observe them? Who are we who look into the darkness?

One of the earliest portraits of an astronomer comes from the pen of a master of English letters, Dr. Samuel Johnson (1709–1784). His sad fable *Rasselas* of 1759 concerns a Prince of Abyssinia sated with pleasure who sets off with his poet friend Imlac and others to seek happiness in the world. They are always disappointed.

Then they meet the astronomer, a man who has spent some 40 years studying the stars. All seems content in the world of the astronomer until suddenly, while observing the satellites of Jupiter, he calls out, "I have possessed for 5 years the regulation of the weather," and then, "The sun has listened to my dictates."

Really? Imlac tries to persuade him otherwise, thinking, with some reason, that the claims are unlikely. But no, the astronomer is certain. He controls all natural forces. Imlac explains that the astronomer has become dominated by the imagination, lost in what Imlac calls 'luscious falsehood': "To indulge the power of fiction and send imagination out upon a wing is often the sport of those who delight too much in silent speculations."

Rasselas is a bleak book. There is no permanent happiness to be found in any way of life. It asks if there is an alternative to madness for those who look too long and on their own into the sky. It wonders what the effects of seeing darkness might be on those who have to look away from their telescopes back to Earth.

These same questions lie behind Thomas Hardy's novel *Two on a Tower*, published in 1882. Hardy (1840–1928) had a grand vision for what the book might be about – "What I have aimed at," he wrote, "[is] to make science, not the mere padding of a romance, but the actual vehicle of romance."

Science as the vehicle of romance – it was a new undertaking, grounded on Hardy's interest in astronomy. His fictional astronomer is Swithin St. Cleeve, whose aim is not a small one. "I hope to be the new Copernicus," he says. "What he was to the solar system I aim to be to the systems beyond."

Indeed, when Tess falls in love with Angel in Hardy's *Tess of the D'Urbervilles*, her feelings "enveloped her as a photosphere.'"

Swithin's observatory sits on top of a tower. There he observes the sky alone. But there he is interrupted by Lady Viviette Constantine. She asks to be introduced to the science. Swithin is reluctant – astronomy is dangerous. Why? He later explains: "Until a person has thought out the stars and their spaces, he has hardly learnt that there are things much more terrible than monsters of shape, namely, monsters of magnitude without known shape."

Although they observe Jupiter and Saturn, Swithin's interest is in the mysteries of deep space. It is there that he takes his new assistant. "The actual sky is a horror," he tells her before they set to work on variable stars.

For Viviette, it is a journey into a new way of seeing. She realizes the grandeur of astronomy, which in turn revives Swithin after his discovery of some new nebulosity is anticipated by another astronomer and his Copernican discovery turns out to be old news. A comet leads him back from illness. Viviette's husband dies. The stage is set for an astronomical affair.

What is Hardy doing in this astronomical love story? It is not his greatest novel, but it was one he cared about. In part, he wants to explore the connection between knowledge of the stars and knowledge of other people. He calls both activities 'observations' – both involve peering into the unknown. He also is interested in the effect of the infinite perspective of deep space upon the little things of Earth. It is something he writes about in his poems, and which Swithin expresses to Viviette: "If…you are restless and anxious about the future, study astronomy at once. Your troubles will be reduced amazingly. But your study will reduce them in a singular way, by reducing the importance of everything. So that the science is still terrible, even as a panacea."

Black holes especially seem to Hardy, and so to Swithin, places wherein our sight might fall and never return, "deep wells for the human mind to let itself down into, leave alone the human body." The perspective of the infinite is for Hardy a belittling one, making our problems, even our consciousness, seem very small.

However, Hardy is also writing about ways of looking upon our own world. Swithin represents the nineteenth-century scientific mind, concerned with the remote, aware that everything will fail. Viviette represents the emotional. She calls the skies beautiful, but Swithin corrects her, saying "They are rather more than that." There is a sense in the novel that, try as he might, Hardy cannot bring these ways of seeing together. Perhaps Swithin has seen too much of the indifferent skies, "that Universe taciturn and drear," as Hardy calls it in one of his poems.

There is another way of looking at what these vast objects might do for humanity. It was expressed by an Editorial in *Nature* at the end of 2011: "The first picture taken of our local supermassive black hole – the most enigmatic and charismatic of all the wonders of the Universe – would surely be one of the defining images of the time. It might even knock everyday trouble and strife from the front pages, and perhaps even, for a while from people's minds."

The vast can make us small. But it can also be a fine distraction, even offering a little comfort.

Chapter 20

Falling in Love

Observation: Saturn, its rings and satellites
Significance: The outer limit of the classical Solar System, rings and rain
Science: Cassini mission, Enceladus, Titan and its environment

Saturn has three distinct but closely bound observing targets: the planet itself, its rings and then its satellites. Each of these is important to the theme of this book. The planet marks the outer limit of the ancient universe, the outermost planet visible to the naked eye. The rings were one of the first great puzzles of the telescopic age, observed by Galileo and still one of the wonders of the Solar System. The moons of Saturn are significant in a quite different way. It is upon one that we find that familiar curse of the astronomer on Earth, rain. The Saturn system, like that of Jupiter, contains rocky worlds where some of the conditions for life may exist.

Observing Saturn and Its Rings

Saturn is a naked-eye object and displays a subtle yellow tinge in the night sky under good seeing conditions. Through a telescope, it turns out to be a delicate flower. It's probably the most beautiful object within range of a small telescope, and it is certainly a sight that converts many a newcomer to the pleasures of astronomy. But treat it badly, piling on too much magnification, and the image can dissolve into a soggy mess.

Saturn is named after the Roman god of agriculture, and its name perhaps derives from the same root as our verb *to sow*. The adjective for the planet is the ugly *Saturnian* and not *saturnine*.

As with all the planets, Saturn appears along the line of the ecliptic. Observing seasons last around 10 months, and the planet lingers in the same constellation for

M. Marett-Crosby, *Twenty-Five Astronomical Observations That Changed the World: And How To Make Them Yourself*, The Patrick Moore Practical Astronomy Series, DOI 10.1007/978-1-4614-6800-4_20, © Springer Science+Business Media New York 2013

long periods. In 2012–2013, the planet is moving between Virgo and Libra. It is worth waiting for opposition, the point at which Saturn is opposite the Sun relative to Earth, since this provides not only a full night's opportunity to see it but also because it is at its closest point to us. Opposition occurs every 378 days, or about 2 weeks later every Earth year.

Fig. 20.1 Saturn, the planet imaged from Finland on 8th May 2007; Camera, Philips ToUCam Pro II – Exposure: 200(3,500)×0,2 s f35; Telescope, Celestron NexStar8i 203 mm f10 SCT; Barlow: Orion Tri-Mag 3×; Filters: Baader IR/UV cut (Credit: Lasse Ekblom, http://www. noba-finland.org/)

Saturn does have belts and zones, but they are more difficult to see than those on Jupiter. This is a function both of distance and of atmospheric haze, making Saturn seem a more peaceable, unvarying world than its giant neighbor. But with enough but not too much magnification – a Barlow Lens helps with this observation – some zones and belts appear. The brightest under almost all conditions is the Equatorial Zone (labeled EZ on observing forms), shining white or pale yellow. With a small telescope, it is often the only feature discernible on the planet's smooth skin.

With more aperture, it is possible to separate the Equatorial Zone into a northern and southern half (EZn and EZs), divided by the very indistinct Equatorial Belt

(EB). It is also in this zone that features can appear like the Great White Spot (identified in 1990) can appear. There seems to be a 57-year cycle for such clouds, so the next might not appear until 2047.

Above and below the Equatorial Zone lie the respective Equatorial Belts for North and South(NEB and SEB). These and the South Polar Region (SPR) are the other features that might appear through a reasonable aperture telescope. It is worth making a sketch or taking notes if any of these become visible – visual report forms are available via the Saturn sections of the British Astronomical Association, Association of Lunar and Planetary Observers or comparable websites.

If you do identify marks on Saturn, note that the equatorial regions where spots tend to appear rotate slightly faster (10 h 14 min) than the higher and lower regions (10 h 38 min). Colors and contrast on Saturn are generally subtler than those on Jupiter. Either a yellow-green filter (Wratten 57) or magenta (W30) improve the overall contrast, while blue or red filters enhance particular features.

Although Jupiter, Uranus and Neptune all have rings, it's those of Saturn that reveal themselves. Searching out details among the rings depends on the angle of their inclination to our line of sight, which varies between 0 ° and + −27 °, when they are displayed in their full glory. The rings are currently opening themselves once more. This has the effect of increasing the planet's apparent magnitude because they reflect light as well. They will be at their fullest extent in 2017.

What is there to see within the rings? The rings are divided by gaps, of which the most apparent is the Cassini Division between the outermost visible or A Ring and the wider B Ring within it. The C Ring lies within the B Ring and is the faintest of the three. There are many more ring sets around the planet, but A, B and C are usually the only ones that can be observed. The Cassini Division, first identified in 1675, is some 3,000 miles across and appears as a clear black space between the two groups of rings.

Much harder to observe is Encke's Division within the A Ring. It lies at the edge of what an 8-in. telescope can achieve. In fact, it's often taken as a good test of the range of an instrument and the quality of seeing. The Keeler Gap, further from the planet towards the outer edge of the A Ring, is even tougher. A 10- or 12-in. instrument is required to capture this.

The Encke Division is named after the German observer, Johann Franz Encke (1791–1865). Encke was honored in this way by American astronomer and founder of the *Astrophysical Journal* James Edward Keeler (1857–1900), who first observed the gap. When *Voyager 1* discovered another gap, it was in turn given Keeler's name.

Observing Saturn's Moons

Like Jupiter, Saturn is accompanied by satellites, some 62 of which have been identified at time of writing. Most of these are very small. Of the five moons that can be observed through a 6-in. instrument, the brightest is Titan, a muddy orange spot shining at a magnitude of +8 or thereabouts. It is the furthest from the planet and its rings. Rhea at +10 lies within Titan about two ring diameters out. The others

– Tethys, Enceladus and Dione – are not easy to spot because of their proximity to the bright rings. Mimas is a real challenge at magnitude 12.1.

Fig. 20.2 The giant satellite Titan behind Saturn's rings, with tiny Epimetheus above them, a combination captured by Cassini's narrow-angle camera on April 28, 2006 (Credit: NASA/ JPL/Space Science Institute)

A really interesting observation is to try to spot Iapetus, which can be almost as bright as Titan but which fades, encompassing a range of magnitudes from 9.5 to 11. How come Iapetus behaves like a variable star? Explorations of the satellite by spacecraft have revealed that the moon has two ill-matched faces, one very bright and the other very dark. Identifying Iapetus at its extremes takes the amateur observer close to one of the strange mysteries of the Saturn system.

To observe Titan is to experience the first observation of a moon of Saturn, achieved by Christiaan Huygens (1629–1695) in 1665. The next four – Iapetus, Rhea, Dione and Tethys – were all identified by Giovanni Domenico Cassini (1625–1712) between 1671 and 1684. Beyond these lie Enceladus and Mimas, both spotted by William Herschel through his 40-ft long instrument in 1789, although he had seen Enceladus through his more modest 20-ft telescope 2 years earlier. To try to observe these two satellites is to experience the skill of Hershel's telescope-making and his patience. Phoebe, identified by W. H. Pickering in 1899, was the first satellite to be discovered photographically

Next time you observe the Saturn system, consider Phoebe. It's too small to spot but is a very interesting object. Probably a captured Kuiper Belt Object, Phoebe may have formed less than four million years after the formation of the Solar System, offering special insights into the very early history of the planets.

It is easy to get confused among the satellites of Saturn. A mnemonic can help – Met Dr. Thip. It lists, in order outwards from the planet, the moons Mimas, Enceladus, Tethys, Dione, Rhea, Titan, Hyperion, Iapetus and Phoebe. This is not, though, any guide to the sizes of the moons, although this is quite easy to remember, for Titan, as its name implies, is the largest by a huge margin over Rhea and then Iapetus.

Probing Saturn's System

Almost everything that follows in this chapter is either a direct result of, or has been profoundly influenced by, NASA's Cassini mission to the Saturn system. Alongside the Pioneers, Voyagers, Galileo and Messenger, Cassini with its ESA lander Huygens belongs in the pantheon of remote explorers of our Solar System – for the science it is still discovering and the beauty it has beheld.

Launched on October 15, 1997, the joint NASA, ESA and Italian Space Agency Cassini spacecraft took nearly 7 years to reach the Saturn system, undertaking gravity assist maneuvers around Venus (twice), Earth and Jupiter. The Earth flyby was controversial because Cassini, like the Pioneer and Voyager deep space probes before it, is powered by three generators that turn energy from decaying plutonium into electricity. NASA was at pains to explain that these were not nuclear reactors and that the alternative power source, the Sun, would require so vast a solar array that the spacecraft would be too massive to launch. In the event, the million-to-one chance of Cassini re-entering Earth's atmosphere never came close to happening. Instead, on July 1, 2004, Cassini entered Saturn's orbit.

The list of Cassini's achievements is immense. It has explored the ring system around Saturn by flying through them, enabling a detailed understanding of their structure and the way they have evolved since Voyager. Similarly different from Voyager's time was Saturn's atmosphere, an unexpected discovery that we will examine further in a moment. Cassini has done extraordinary science among the moons, examining Enceladus, Dione, Iapetus and Phoebe but most especially Titan. On November 27, 2004, the Huygens probe touched down on the surface of Titan, the first landing achieved by a human instrument in the outer Solar System.

In April 2008, the Cassini mission was extended for 2 years and renamed Equinox. In February 2010, with the spacecraft still functioning well, a further 7 years of scientific life were granted to it. Cassini Solstice, as it is now called, will continue to operate until 2017.

We will explore four of Cassini's scientific achievements in more detail. Each of them enrich our observations of the planet, its rings and two of its principal satellites. Each augments our sense of the beauty of the system. Each is also part of the transformation of our Solar System that Cassini is achieving.

Saturn in the Spring

Cassini has had time to watch the atmosphere of Saturn through a change of seasons. The poles of the planet alternate between 15 years of darkness and 15 of sunlight. If this was not change enough, Cassini has also been able to measure the effects of the shadows cast over the atmosphere by the ring systems. When the angles reduced the shadow between 2004 and 2009, the stratosphere in Saturn's northern mid latitudes warmed by 6–8 K.

The color of the atmosphere has changed as well. As winter became spring, the blue of the northern hemisphere became the more familiar yellow, demonstrating a close relationship between season, temperature and color.

Cassini has been able to penetrate the inner atmosphere. Although the patterns of its deep rotations are still unknown, the probe's Visual and Infrared Mapping Spectrometer (VIMS) instrument was able to observe clouds beneath the surface haze by mapping silhouettes against the glow of the planet's own heat.

A part of this hidden interior is captured in this remarkable image, constructed from 25 images taken over 13 h in a false-color mosaic. It reveals Saturn in nighttime and daytime conditions, puncturing the haze to show a dense array of narrow bands and cloud features within the outer shell of atmosphere. The deep clouds are probably formed from ammonium hydrosulphide. The rings, meanwhile, partly smothered by the planet's shadow, shine brilliant blue.

Fig. 20.3 This image was acquired by Cassini on Feb. 24, 2007. Data at wavelengths of 2.3, 3.0 and 5.1 μm were combined in the *blue*, *green* and *red* channels of a standard color image, respectively, to make this false-color mosaic (Credit: NASA/JPL/University of Arizona)

There will be more to come. The extended Cassini Solstice mission will enable scientists to observe the progress of summer across the planet's north, as well as to take further measurements of a wavy atmospheric pattern in the equatorial areas, "candy cane-like," according to NASA, with a "striped, hot-cold pattern."

This Most Deceitful Star

The rings of Saturn made Saturn a perplexing observation to the first generation of telescopic observers. Galileo first thought that Saturn comprised three objects, the planet and two moons. "Triple-bound" was how he described them. But late in 1612, the planet had changed, now "solitary, perfectly round and clearly defined as Jupiter."

Worse was to follow. In 1616 he reported observing: "The star of Saturn whose two companions are no longer two small, perfectly round globes as they were before, but are at present much larger bodies, and no longer round."

What was causing these alterations? Galileo remained interested in Saturn until the end of his life, but it was left to his successors to probe the planet in more detail. 'Handled' (the Latin word they used was *ansa*) was how many described it, with a central sphere and two protruding sections. The changes fascinated astronomers such as Gassendi, Hevelius and Riccioli, and many theories circulated as to the cause. Hodierna thought that they were looking at "raging oceans, enclosed by land"' Christopher Wren noted this enthusiasm for Saturn, which was: "Proposed as the greatest test of skill. This is the target upon which they aim their artfully strengthened vision and they strive to bind this most deceitful star with the laws of a particular hypothesis."

Early in 1656 Christian Huygens (1629–1695) turned his new telescope upon deceitful Saturn. In March he circulated a prediction that the 'handles' of Saturn, invisible at the time, would reappear in April. He accompanied this with a 58-letter anagram, and those who could not decipher this had to wait. It was 3 years before Huygens explained the mystery in his book, *The System of Saturn*: "He is surrounded by a thin flat ring which does not touch him anywhere and is inclined to the ecliptic."

Huygens' anagram was this: Aaaaaaccccdeeeeghiiiiiiillllmmnnnnnnnnnnooooo ppqrrsttttttuuuu. The meaning seems not to have been obvious.

A ring fitted the evidence. The observations were affected by the changing inclination of these rings as seen from Earth. But there were problems caused by the solution, not least of which was how a solid ring could exist anywhere at all.

It was James Clerk Maxwell (1831–1879) who explained the nature of the rings. He did so not as a result of observation but by constructing models of the planet. "I am still at Saturn's rings," he wrote in December 1857. "At present two rings of Saturn's are disturbing one another. I have devised a machine to exhibit the motions...for the edification of sensible image worshippers." The model was built the following year. It enabled him to explain in his *On the Stability of the Motion of Saturn's Rings* (1869) that Saturn's was "a ring, the parts of which are not rigidly connected." Every separate particle acted as a separate satellite, orbiting independently. The rings, therefore, were largely made from dust.

The Voyager flybys in November 1980 and August 1981 confirmed Clerk Maxwell's theory. The rings were indeed formed from particles, some very small but others larger, as much as tens of meters across. Cassini's detailed surveys have shown that these particles are made of, or at least coated in, crystalline ice. It has shown how the material in the rings form clumps under the influence of Saturn's satellites, providing evidence in support of a theory first advanced by Daniel Kirkwood (1814–1895) in 1866. Kirkwood proposed that the Cassini division within the rings was caused by resonance with a moon. The role of these shepherd moons within the rings has now been observed in detail.

The intimacy of moons to rings is demonstrated in the outer G-Ring. Cassini scientists, examining a set of pictures acquired over the course of over 2 years of the probe's work, found a tiny moon where previously there had been none. The G-Ring had seemed odd in that it had no associated satellite. The moonlet, now named Aegaeon in place of its preliminary designation S/2008 S 1, is thought to be about half a kilometer across. Meteoroid impacts and the clump-collision process are thought to be feeding the tenuous ring.

The rings of Saturn are dynamic environments. In April 2012, Cassini images showed the activities of so-called 'mini-jets' within the F-Ring. Objects were moving in packs and causing collisions. It was reported by NASA that the F-Ring was 'a bustling zoo of objects' – understanding the mechanisms involved here may well provide insights into the dusty rings from which planets are formed around young suns.

The moons were originally named after Saturn's mythological siblings, the Titans and Titanesses, giants who had ruled the heavens before Jupiter conquered them. The choice was made in 1847 by Sir John Hershel, who pointed out that Saturn's children would not be suitable, as in mythology Saturn ate them. Gallic, Inuit and Norse names have now augmented the list in order to distinguish different orbit inclination groups with respect to the ecliptic. Retrograde satellites except Phoebe carry the names of Norse giants, giantesses and wolves (Suttungr, Skathi and friends), while prograde satellites are divided between Gallic giants such as Albiorix and Inuit giants and spirits beginning with Paaliaq. Aegaeon marks a return to the Greek. He was a giant boasting a 100 arms.

Enceladus Lives

The largest and outermost ring of Saturn, the E-Ring, has a different origin. It is almost certainly being fed by eruptions from Enceladus. On its third flyby of the satellite on July 14, 2005, Cassini's Cosmic Dust Analyzer (CDA) measured directly the ice grains pouring from Enceladus' south pole. Cassini had explained the source of the ring. It had also pierced the secrets of a volcanically active world.

Enceladus can claim many distinctions. It has the highest albedo of any Solar System body as light is reflected off its bright-white icy surface. The Hubble Space

Telescope discovered it was emitting hydroxide (OH), and Cassini detected it was also releasing oxygen. The July 2005 flyby provided conclusive evidence of activity within the moon, with thermal emissions identified near the south pole that seemed to be associated with fresh 'tiger stripe' fractures in the ice.

What is erupting from within Enceladus? Using the Imaging Science Subsystem (ISS) instrument to measure the brightness of the solid particles in the plumes, it has been possible to establish that Enceladus is venting ice grains and water vapor, together with carbon dioxide and other gases. The Cosmic Dust Analyzer has further identified the presence of salts, including common table salt sodium chloride. Deeper penetration into the plumes during the July 2008 and subsequent flybys have confirmed this, and there will be further close approaches to the satellite during the Solstice phase of the mission. At time of this writing, Cassini is preparing for its nineteenth encounter with the world.

Enceladus joins the exclusive society of volcanically active worlds. Its geological activity is cryovolcanic (the *cryo-* prefix derives from the Greek word for frost), and the power that drives the eruptions is tidal, drawn from the 2:1 resonance between Enceladus and its neighbor Dione. At present, the activity is confined to the south polar region of the moon, where it is associated with four linear cracks some 500 m deep, 2 km wide and 130 km long. Named the Alexandria, Cairo, Baghdad and Damascus Sulci (*sulcus*, the single form of *sulci*, is Latin for a furrow or trench), these 'tiger stripes' are overlaid near the active spots with smooth deposits.

Fig. 20.4 Enceladus, taken in visible light with the Cassini spacecraft narrow-angle camera on Nov. 6, 2011 (Credit: NASA/JPL-Caltech/Space Science Institute)

Ice rules Enceladus. The presence of so much water ice and the evidence of internal heat driving the plumes creates the possibility of liquid water existing beneath the icy crust. Some models for the internal structure of the moon predict a vast subsurface sea enveloping the whole globe. The presence of sodium chloride in the vents suggests that the subsurface water is in contact with the moon's rocky core. Enceladus must therefore be a significant candidate for the future search for life – this is one of the many questions that the Cassini mission will hand down to the next generation of (presumably unmanned) explorers.

A Landing on Titan

Titan is the furthest point where human ingenuity has left a footprint. First observed by Christian Huygens in 1655, Sir John Hershel declared its name to be Titan because of its size. It is the largest moon of Saturn and the second largest in the Solar System after Jupiter's Ganymede, substantially larger than both our Moon and the planet Mercury.

Titan was also known before Cassini to be the only satellite in the Solar System with a substantial atmosphere. This was detected for certain in 1944 by Gerald Kuiper, using the new 82-in. telescope at the McDonald Observatory.

Much more was learned about Titan after flybys in 1979 (*Pioneer 11*), 1980 (*Voyager 1*) and 1981 (*Voyager 2*). It was visually disappointing, shrouded entirely by an orange haze, but they measured the surface temperature and pressure and were able to confirm Kuiper's atmosphere, primarily nitrogen but also laden with methane.

Huygens was a European Space Agency 2.7-m clam-shaped probe, protected by a hard shell and designed to survive descent and landing and then conduct 3 h of science experiments with six instruments. Its goals were to identify and measure the contents of the atmosphere and wind speeds but above all to find out what the surface of the planet was like. For a start, was Titan solid or liquid? Would Huygens sit on rock or have to float?

"There will only be one first successful landing on Titan," NASA Associate Administrator Al Diaz said. Was this going to be it? Dispatching a lander through a thick atmosphere onto an unknown surface was no small undertaking, and there were long minutes when it was not clear whether Huygens had made it or not. On January 14, 2005, the scientific world waited for Cassini to receive its first signals from Huygens – four Earth-based telescopes had watched for a fireball and not seen one, so that at least was something. Then news came from Australia, where the Parkes radio telescope had detected Huygens' carrier signal. Minutes later it was confirmed that Huygens had landed between 1:45 and 1:46 p.m., Central European Time. Hours passed before Cassini turned its attention away from the probe and started relaying information back to Earth. Finally, applause exploded around Huygens mission control – the data had arrived. Some 350 images were collected, along with a wealth of other information.

And shown here is Titan on that day, one of the most extraordinary pictures yet collected by a remote probe.

Fig. 20.5 The surface of Titan Credit: ESA/NASA/JPL/University of Arizona

What does it reveal? Huygens landed in a dry riverbed where once had flowed not water but methane. The surface was littered with dirty water-ice rocks. There was evidence of methane rain having fallen in the past.

It now seems clear that methane plays a role on Titan analogous to that of water on Earth. With a surface temperature of −179 C, any water that there is will be forever sealed as ice, but methane precipitates from clouds, forms valleys, river systems and even lakes. The icy chunks imaged by Huygens showed signs of erosion by liquid, and Cassini radar images have revealed methane lakes. One of these, the Kraken Mare, has a shoreline of coves, inlets and islands.

How often does it rain on Titan? The liquid patches identified by Cassini are all at the poles, while conditions near the equator are generally arid. The landscape here is dominated by huge and towering dunes where the 'sand' is either pulverized ice or organic material that has fallen from the atmosphere. But observations in October 2010 suggest surface darkening after a cloudy period at the Belet dune field, almost certainly caused by a large methane storm. Although it is accepted that rain on Titan is much rarer than it is on Earth, the implication of the Cassini data is that liquid fell quite suddenly in what was, in effect, an April deluge.

It has also been possible to detect the other end of the methane cycle on the moon, evaporation. Studies at the Ontario Lacus have confirmed a receding shore-line and a 1 m per year decline in levels of liquid. This is, in short, a climate.

One question that arises is the origin of the methane, which is destroyed by sunlight and must therefore be constantly replenished. There may well be volcanic features on the moon – the Hotei Regio and western Xanadu are two likely loca-tions – and something must have caused the uplift of Titan's mountains.

Is Titan a place where life might have developed? It has been suggested that when we look at Titan, we are seeing a frozen moment in Earth's own history and that, in particular, the precipitation of organic molecules from the surrounding smog onto the surface may have been the mechanism by which such molecules were delivered to Earth's oceans. Titan may occasionally have liquid water pro-duced by cryovolcanism or after large impacts, and organic chemistry is taking place in the upper atmosphere. There may also be a subsurface environment of ammonia and water, an extreme locus for life, but not impossible. The fact that there is another world than ours where sunlight can glint off a liquid lake, here the Kraken Mare, will make us want to explore Titan more.

Two Cultures or One?

The number theorist G. H. Hardy (1877–1947) wrote in his *A Mathematician's Apology* that: "A mathematician, like a painter or a poet, is a maker of patterns. If his patterns are more permanent than theirs, it is because they are made with ideas."

If this seems a strange likeness, mathematician compared to poet, it is because we have become used to the idea of a disconnection between the sciences and the arts, the one concerned with chill, harsh facts and the other with beauty and emo-tions. These were labeled the Two Cultures in 1959 by one of Hardy's friends, the British scientist turned writer of fiction C. P. Snow (1895–1980). He set the chal-lenge between science and the arts in these terms: "I have been provoked and have asked the company how many of them could describe the Second Law of Thermodynamics. The response was cold: it was also negative. Yet I was asking something which is about the scientific equivalent of 'Have you read a work of Shakespeare's?'"

A bitter personal assault on Snow came from the literary critic F. R. Leavis, who described Snow as "intellectually as undistinguished as it is possible to be."

The Two Cultures is shorthand for the distinction between two streams of educa-tion. It expresses the fairly obvious point that the sciences and humanities do things in different ways. In the arts, everything is always open to interpretation. Think of how many portraits of one person can all be different and all be true – the same is not the case for the absolute magnitude of distant stars.

Setting aside some of the weaknesses of Snow's own argument – in its first form, especially, he seems to equate human happiness only with technological progress – there is a relationship between the sciences and the arts that cannot be shrunk into a single phrase. Science and the humanities do not stare at one another across a great divide. Aside from the long lists of scientist-artists and artists with a keen understanding of science, creators of science-in-fiction as well as science fiction, both the sciences and the humanities are seeking the same end, truth, and both require imagination, creativity and experiment. The difference lies in the kinds of questions asked and the kinds of answers sought. This is the scientist-novelist Alan Lightman's diagnosis:

> Scientists work on questions with answers....That answer may take ten years to find, or a hundred, but an answer exists. By contrast, for artists the question is more interesting than the answer, and often the answer doesn't exist. How does one answer a question such as 'What is love?' or "Would we be happier if we lived to a thousand years?' (*Nature* Volume 434, March 17, 2005, p. 299)

Both science and the humanities respond to beauty. This the common language. G. H. Hardy says of his discipline that: "The mathematician's patterns, like the painter's or the poet's, must be beautiful.... Beauty is the first test; there is no permanent place in the world for ugly mathematics."

Saturn as envisioned by Cassini represents why the separation of the sciences and the humanities cannot be right. Science gets Cassini there; it asks the questions and it makes sense of the answers. But then the wonder of those answers and the images taken in pursuit of the science are a feast for the imagination. They also inform our moral sense of where as a species we find ourselves in the universe. So maybe the answer to Snow's Two Cultures lies in perceptions like the following. Here is Humphrey Davy (1778–1829), chemist and president of the Royal Society in London: "The perception of truth is almost as simple a feeling as the perception of beauty; and the genius of Newton, of Shakespeare...are not very remote in character from each other. Imagination, as well as the reason, is necessary to perfection in the philosophic mind."

Chapter 21

Into the Deep

Observation: M77, Virgo, M60, M87, M88, M91
Significance: Measuring deep space, Olber's paradox, dark matter
Science: Hubble V1, Seyfert galaxies, red-shift, the evidence for dark matter

In this chapter we penetrate the deep places of space. Our observations take us to far-off galaxies, including some of those in the Virgo supercluster, one of the richest areas of the night sky. We look at how resolving the strange fact of a dark sky illumined by so many stars leads to a fundamental property of the universe, its expansion. We then try to peer at the darkest thing of all, dark matter.

The Most Important Star in the Sky

It is always worth returning to the Andromeda Galaxy, M31. It is, as we've established, a naked-eye observation that takes us beyond the Milky Way and to the shore of a separate island universe.

Edwin Hubble himself spent months in 1923 scanning Andromeda, using the 100-in. Mount Wilson telescope to search deep inside the spiral arms. On October 5, a night of poor seeing, Hubble took a 45 min exposure of what he thought at first was a nova, but which he later identified as a Cepheid variable, labeled V1.

Why did this star matter? As we saw in the chapter on variable stars, Cepheids are reliable markers of distance within the Milky Way. Hubble was able to use this to measure a distance to Andromeda of one million light years. He later reduced this as more Cepheids were found, and proposed that the galaxy was 900,000 light years away.

M. Marett-Crosby, *Twenty-Five Astronomical Observations That Changed the World: And How To Make Them Yourself*, The Patrick Moore Practical Astronomy Series, DOI 10.1007/978-1-4614-6800-4_21, © Springer Science+Business Media New York 2013

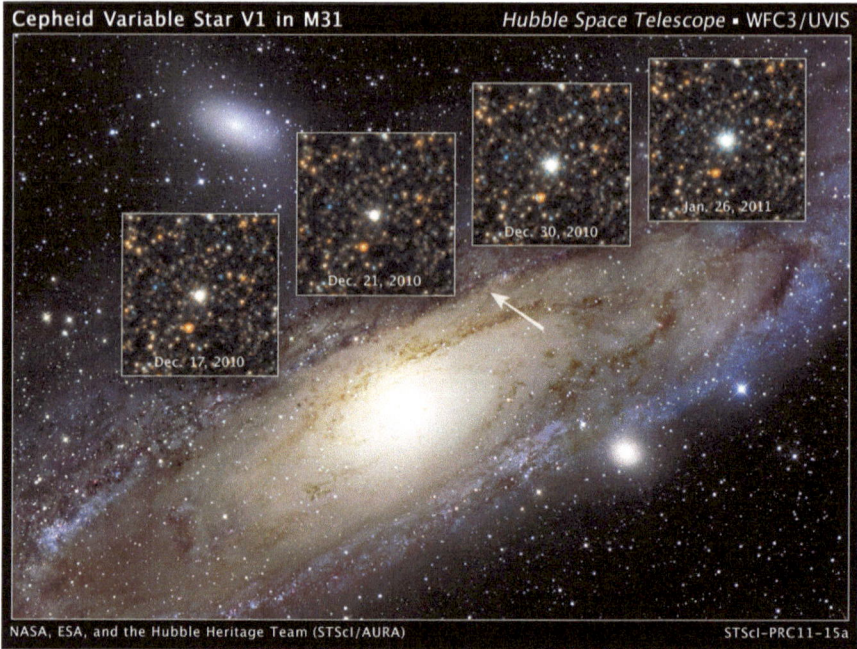

Fig. 21.1 Hubble views the star that changed the Universe (Credit: NASA, ESA, and the Hubble Heritage Team (STScI/AURA); Acknowledgment: R. Gendler)

In December 2010 and January 2011, the Hubble Space Telescope repeated his observations, focusing once again upon V1, "the most important star in the history of cosmology," according to Dave Soderblom who proposed the investigation. The resulting image, as seen here, shows Hubble's famous star acting as the Cepheid lighthouse, guiding astronomers towards understanding the scale of the universe. With the help of a magnitude 19 eclipsing binary in the galaxy, we now set Andromeda some 2.52 million light years from Earth. Hubble was more right than he imagined. This gives us a yardstick for the abyss of deep space.

Blue Spirals and Active Cores

Our first observation takes us back to the constellation Cetus the Whale and the Wonderful Star Mira (Delta Ceti), which lies close alongside the deep-sky galaxy M77. The Earl of Rosse, observing this from Ireland through his legendary 72-in. reflector Leviathan telescope, spoke of seeing a blue spiral here, and while Rosse's color is elusive, the spirals certainly are not.

M77 is a beautiful observing target and easy to find, since it appears adjacent to Mira. This is an illusion, however. M77 is at least 46 million light years away, maybe much more. With a magnitude of 8.9, it is well within range of amateur 'scopes, although it can appear low on the horizon for some northern hemisphere observers. The galaxy's most obvious feature is its core, which appears much like a star and is, importantly for the science that follows, strikingly bright. Through an 8-in. instrument, M77 can seem deceivingly comet-like, a trick caused by a greater density of stars in the northwestern part of its halo. A 12-in. aperture will reveal the inner spiral arms, and if the seeing is steady, some structure of these and even the galaxy's outer limbs.

Fig. 21.2 M77, imaged with Celestron 9.25" F/10 SCT, ST8XE, L: 60 m, RGB: 10 m, 2X binned, from Chandler, Arizona (Credit: Richard Jacobs, http://www.azastronomy.com/)

The image here shows the kind of resolution of M77 that is possible with amateur instruments.

Having got close to the galaxy, pan out using a wide field eyepiece. With M77 towards the bottom of the image, you will glimpse its northern neighbor NGC1055. In contrast to the Messier object, this galaxy is side-on to our line of sight, and a decent aperture will reveal the dust lane on its outer rim.

M77 is immense. It is estimated to contain some one trillion solar masses and to span as much as 170,000 light years, making it one of the largest deep-sky objects in the Messier catalog. It is also a scientifically significant observation, for that compact, star-like core is a clue to the intense activity at its heart. Studies of spiral galaxies in the early twentieth century using nascent spectroscopic techniques indicated that some galaxies displayed unusual and strong emission lines. These were first identified for M77 in 1908 by Edward Arthur Fath (1880–1959), working at the Lick Observatory. M77 was also one of the galaxies studied by Carl Keenan Seyfert (1911–1960), who demonstrated that these emission lines were caused by matter approaching and receding from us at high velocities. Galaxies exhibiting this kind of behavior were named Seyfert galaxies in honor of this discovery.

What is going on in these galactic cores? Seyfert galaxies have active galactic nuclei (AGN), and their key characteristic is a variable excess of radiation, especially in the infrared. This radiation comes from a tiny region when compared to the whole galaxy, and it is now believed that Seyfert's emission lines point to matter falling into supermassive black holes in the galactic heart.

A good number of Seyfert galaxies can be observed without too much difficulty. There is a rewarding cluster in the lower part of Ursa Major beneath the Plough asterism. The bright star Phecda (Gamma Ursae Majoris, the lowest star in the group in the direction of Leo) is a useful guide. A line between the first star of the Big Dipper's handle and Phecda stretches away from the asterism towards NGC3718, a magnitude 10 face-on barred spiral with exactly the sort of bright nucleus that indicates intense activity. The same field of view as this galaxy also allows a glimpse of NGC3729. Meanwhile if you move from Phecda down to the K-type orange-yellow Chi Ursa Majoris, you are not far from NGC4151, sometimes called the Eye of Sauron. A recent observation by the Jacobus Kateyn Telescope at La Palma has shown it to harbor a supermassive black hole at its heart. At a mere 43 million light years from Earth, it is one of the nearest actively growing black holes to us and offers a special chance to study the workings of such a galaxy.

Why Is the Night Sky Dark?

These galaxy observations are not easy. The darkness seems to overwhelm us as we hunt through space for far-distant objects. Raising the question – why is the sky dark? If the universe is without limits and contains an uncountable multitude of stars, then any straight line of sight from Earth must at some point encounter a star. It seems to follow from that that the whole sky should be ablaze with starlight.

The problem of the dark sky is often named Olbers' paradox after Heinrich Wilhelm Matthaus Olbers (1758–1840), whose paper of 1823, *On The Transparency of Space*, opens with an appeal to the infinity of the sky. "Did the late keen-eyed Herschel penetrate to the outer limits of the universe?" Olbers asks. He did not. Indeed, "endless space itself is sprinkled with suns and their accompanying planets and comets." The only way that Olbers could explain the paradox was that space was not transparent, and that starlight was absorbed by the interstellar medium.

We can probably have a better stab at resolving Olber's paradox now.

In 1913, Vesto Slipher (1875–1969), working at the Lowell Observatory in Arizona, wrote of the Andromeda galaxy: "'We may conclude that the Andromeda Nebula is approaching the solar system."

This was an astonishing insight, which Slipher developed for a presentation to the American Astronomical Society in the dreadful month of August 1914. Far from the drumroll of war in Europe, Slipher demonstrated spectrographic studies of 40 nebulae and clusters and identified their velocities. Some were moving away from Earth and others approaching. Using spectrographs laboriously constructed by Milton Humason (1891–1972), Slipher and later Edwin Hubble pioneered the study of red-shift as evidence for the motions of galaxies.

Red-shift and its analog blue-shift refer to the way that spectral lines observed in galaxies are found to have moved towards longer or shorter wavelengths. Using just 24 galaxies, Hubble showed that the universe was expanding, with every point moving away from every other, the relationship between velocity of recession and distance bound by what has come to be known as the Hubble constant (H_0). Hubble himself originally estimated H_0 at about 160 km per second per million light years. Controversy has surrounded this value ever since, and it has risen and fallen with new theories and measurements. When the Hubble Space Telescope was given its patron's task, it helped to establish a current best estimate for H_0 at 71 (+ − 7) km per second per million light years.

It is hard to overvalue the effect of this conclusion. If we use the word revolution to describe first the establishment of a Copernican universe and second the achievement of Galileo and the telescopic astronomers following him, then the discovery of an expanding universe must count as the third great revolution in astronomy.

It was also used to solve the dark night paradox. Writing in 1952, the cosmologist Sir Hermann Bondi (1919–2005) explained that: "If distant stars are receding rapidly, the light emitted by them will appear reddened on reception and hence will have lost part of its energy." The darkness of the night is a sign of the accelerating universe.

Let us turn now to the richest source of deep-sky observations in the night sky, the supercluster in the constellation Virgo.

Detailed studies of galaxy clusters can be said to have started with the 1958 catalog of 2,712 clusters compiled by George Ogden Abell (1927–1983) using the 48-in. telescope at Mount Palomar. Abell established some of the basic properties of clusters, including their degree of symmetry, their distribution, their richness (how many galaxies they contain) and their size. He found that most galaxy clusters were contained within a radius of about 2 megaparsecs, often referred to as the Abell radius.

The parsec is a unit of measurement used in deep sky astronomy derived from the trigonometry of parallax. It avoids the huge numbers that would be involved in sticking with AU or light years. One parsec is equivalent to 3.26 light years or 3.09×10^{16} m. A megaparsec is a million parsecs or 3.09×10^{22} m.

Examining clusters gives us a sense for the universe as a whole. Since they contain so many galaxies and stars of different ages and states, they are cumulatively a study of the average across the universe. They give us clues about the universe's basic composition, therefore its origins and perhaps also its ultimate fate.

With these grand ends in mind, let us look out into the deep. Virgo is the second largest constellation after Hydra and can be found in two ways. The first and easiest is to follow the sweep of the 'tail' in the Plough asterism, which arcs through the sky through Böotes and then to the bright star Spica, Alpha Virginis. More interesting, perhaps, is to start with the backwards question mark asterism in Leo and blue-white Regulus, a main sequence star that rotates so fast that it is flattened by the effect of speed. This is not unknown among stars, and we have met the phenomenon before. The problem is that stars, like people, are expected to slow down over time, and Regulus is too old to exhibit this sort of behavior. The explanation probably lies in mass exchange and other interactions with either Regulus' wide companion Alpha Leonis B or, and this seems more likely, with a faint white dwarf that is bound to it.

But these are the concerns of our own galaxy. Move away from Regulus in the direction opposite to bright Sirius and two guide stars lead us on. The upper one of these is Arcturus, Alpha Böotes, a deep yellow star tending to orange, lovely to observe. Below Arcturus lies Spica, which like Regulus is a B-type star. With Spica at the bottom of the field of view in binoculars, the star at the top towards Regulus is Gamma Virginis, Porrima, a double star that can with care be separated in an amateur 'scope. Porrima is made up of two stars of near identical magnitude +3.6, orbiting each other over 169 years. At time of this writing they are very close, but they are drifting apart, and separating them will become easier in time. Lying close to the ecliptic, Porrima is a good target for planetary occultations as well.

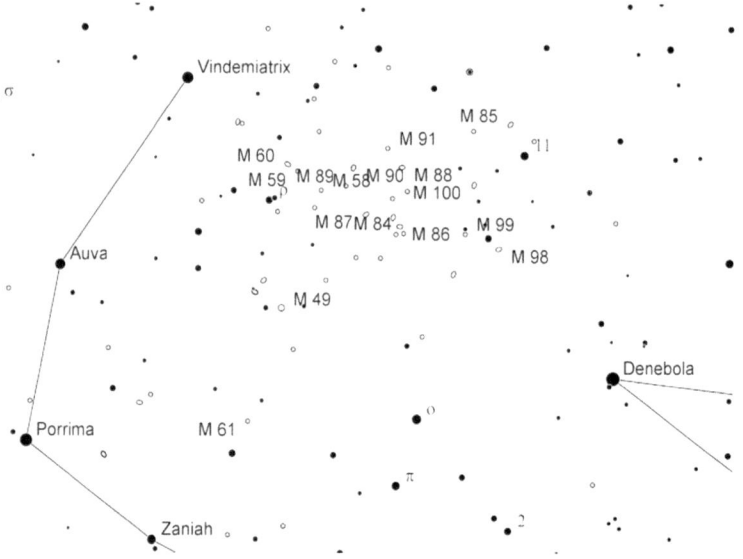

Fig. 21.3 Part of the constellation Virgo. Cartes du Ciel

Continuing from Porrima there are two stars above it that form the boundary of our deep sky observing. The first is Delta Virginis, a golden red giant, and the upper is G-type Vindemiatrix, the Grape Gatherer, more prosaically Eta Virginis.

The names Porrima and Spica are of Latin origin. Porrima preserves the name of an ancient deity of prophecy, while Spica is a short form of *spica virginis,* Virgo's ear of wheat.

Become familiar with Vindemiatrix. The area between this star and the edge of Leo contains a remarkable cluster of galaxies known as the Virgo Supercluster, of which the nearest to the star, and one of the brightest in the group, is M60, with a magnitude of +8.8. M60 even has its own companion galaxy, NGC4647, but they are a tricky pair to capture because with any decent aperture, M60 floods its neighbor with light. M60 itself is a little featureless, with a stellar core that fades quickly into the darkness. NGC4647 meanwhile is often a smudge and not much more, maybe two-thirds of the diameter of M60.

Fig. 21.4 M60 and NGC4647, imaged with a 32-in. Schulman telescope (RC optical systems) and SBIG STX 16803 CCD camera in December 2011, Mount Lemmon SkyCenter, Arizona (Credit: Adam Block/Mount Lemmon SkyCenter/University of Arizona, http://www. caelumobservatory.com)

From M60, it is a small jump towards Leo to the great elliptical galaxy M87. The galaxy's bright core dominates, and only at the top end of what most of us can achieve does its halo become visible. M87 is famous for the jet that roars from its heart, throwing material some 250,000 light years into space. Images of this were captured in the 100-in. Hooker telescope at Mount Wilson by Otto Struve and then in 1992 and 2009 by the Hubble Space Telescope. But it takes exceptional equipment and skies to glimpse it for oneself, and there is much else to see in the immediate neighborhood, including both M84 and M86, two blurred marks through binoculars that resolve in larger apertures into a pair of near-twins.

The Matter of Darkness

It will not have escaped your notice that some of these deep sky objects are tricky to find. If you have the sense that they should be more luminous, you're right. This is not, though, about our convenience. The brightness, or lack of it, of galaxy clusters takes us to an observation that is changing the way we understand the universe.

It was in the 1930s that Fritz Zwicky (1898–1974) attempted to determine the masses of clusters. He made use of the virial theorem, which states that the magnitude of gravitational potential energy of a bound system is twice the total kinetic energy, to establish the relationship of velocity to mass. If the velocity could be measured, the mass could be determined. He applied this to the Coma Cluster while Sinclair Smith (1899–1938) did the same for Virgo. At the same time, they examined the speeds of galaxies in the clusters.

Zwicky was a visionary cosmologist, who among other achievements was the first to coin the word supernova to distinguish an exploding star from an eruptive nova. But insightfulness did not make Zwicky easy to work with. His aggressive attitude towards colleagues has remained at least as legendary as his ideas.

The results were, in Zwicky's words, "somewhat unexpected." This was a mild way of expressing the startling fact that the clusters were a lot heavier than their luminosity indicated. There was either a lot of light missing or a lot of mass within the cluster that was not giving out any light at all. Also, the galaxies were orbiting so fast that their collective mass, as measured by the luminosity of the cluster, was not large enough to stop the galaxies from spinning away.

It was from these two findings that there emerged the concept of dark matter. It was the missing mass. It is now estimated that the matter with which we and galaxies are made of, known as baryonic matter, comprises just 4.6 % of all matter in the universe, with the non-baryonic divided between dark matter (23 %) and dark energy (72 %). The nature of these dark presences is immensely difficult to explore. Non-baryonic matter interacts either very little or not at all with the baryonic. Most studies of dark matter are today undertaken either by the Zwicky method or with gravitational lensing and the study of Einstein Rings and Crosses, but while there have been some claims at direct detection of dark matter in some underground laboratories, the habits and forms of dark matter remain theoretical rather than proven, and there are some alternative models that claim to explain the missing mass and the velocity of galaxies in clusters.

This famous image of the cosmic microwave background does give a further clue as to how the universe works. About 380,000 years after the Big Bang, the temperature of the universe fell sufficiently to allow atoms to form. This released a huge amount of light energy, stretched since then by expansion into microwave wavelengths.

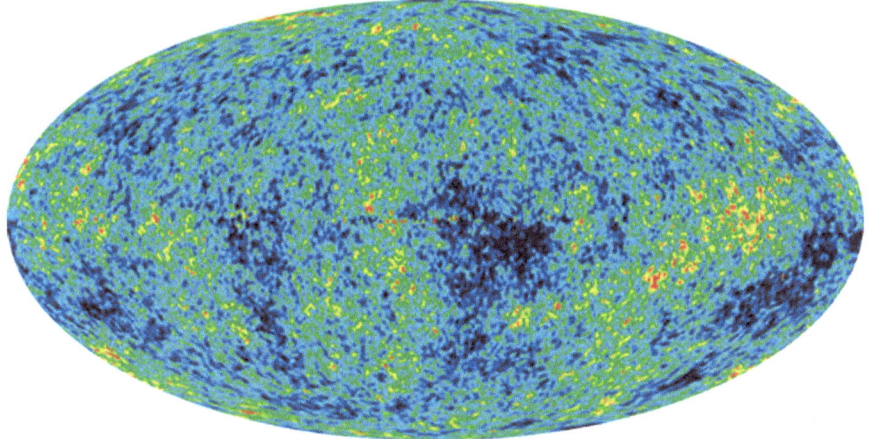

Fig. 21.5 The WMAP image of the cosmic microwave background (Credit: NASA/WMAP science team)

It is not completely uniform. The image, created out of 7 years of data collected by NASA's Wilkinson Microwave Anistropy Probe (WMAP) mission, reveals patches that are slightly hotter than others. It seems to show that dark matter came together in clumps before normal matter, and the pattern of spots indicates that, for every 1 g of luminous material in the universe, there is about 4 g of dark matter gathered into haloes around galaxies and clusters.

This discovery is the threshold to another cosmological revolution. If the telescope has, in the hands of Galileo, Herschel and many others, brought us the astronomy of the visible, studies of dark matter promise a new astronomy, this time of the invisible. When we struggle with the dimness of some long-sought deep sky object, we are actually making an observation of profound significance for the universe.

Back to Virgo

Our final set of observations takes us back to the Virgo supercluster. Our guide stars are Vindemiatrix and the edge of Leo, marked by the bright A-type star Beta Leonis or Denebola, the Lion's Tail. The truth is that this is an observation that can't ever be finished. There are so many tantalizing deep sky objects on the edge of telescope power and averted vision, with at least 2,000 galaxies in the Virgo area alone.

Spend time with M88. If you track from Denebola towards Vindemiatrix using binoculars, you will be able to see the tiniest of dots that is the galaxy. Locating it in the telescope is a challenge, but keep the Leonine star as your place of refuge if you get lost. M88 is a barred spiral galaxy and probably a near analog to our own Milky Way.

Fig. 21.6 M88, image captured on 19 June 2009; Camera: Meade DSI Color II; Exposure: 14 m 30 s (17×30 s) RGB+(14×30 s) L; Focus Method: Prime focus; Telescope Aperature/ Focal Length: 203×812 mm; Mount: LXD75; Telescope: Meade 8" Schmidt-Newtonian (Credit: Ed Sunder, http://flintstonestargazing.com)

Tracking east from M88 leads to faint M91, another spiral galaxy but a difficult one to see well. M91 is an obstacle to many who want to get through the Messier challenge. Though listed as a magnitude 10 object, it conceals itself remarkably well. The only way to glimpse it is by averted vision, and even then it is more of an impression than a certainty.

But M91 is nevertheless important as a stepping stone to M90. This is tough without go-to software, and a 12- to 14-in. telescope is the only way to see its spirals. From M91 move southeast towards the heart of Virgo, and M90 will be a pallid creature, notwithstanding the suggestion that it might be substantially closer to us than other galaxies in the cluster. From M90, it is a short move further southwest to the very bright M89.

M89 is not, in truth, a fascinating observation. It is a small, compact nebula with two quite pretty stars on either side of it. But it is significant, because it was in M89 that David Malin detected the remnants of an ancient galaxy encounter. He identified a jet some 150,000 light years long, which is either the relic of a consumed galaxy or the residue of a brushing encounter with a galaxy that still survives.

We cannot of course see these galaxies in motion. Observations in Virgo tend to reinforce the idea of galaxies as island universes, alone in the dark. But we can glimpse moments in their lives as galaxies collide. Two conditions seem to be essential for such encounters: the galaxies must of course be close enough in space, but they must also be traveling slowly enough, relative to each other, for gravitational bonds to form. Inevitably, therefore, these events are more likely in galaxy groups.

How then can we make sense of the stages of galaxy encounter? Computer simulations have played an important role in the science of collisions since brothers Alar and Jüri Toomre first made use of them in the 1970s. Their modeling of galaxies led to the creation of the Toomre Sequence, and images from space-based telescopes have seemed to confirm the outlines of a process. Galaxies approach and then form tidal tails before the centers merge. These tails persist and then dissipate so that the new galaxy takes on an elliptical but disturbed shape. Quiescence, the end of the collision when one galaxy has absorbed the other, can take some 500 million years to reach.

Our own Milky Way has experienced somewhere between 5 and 11 minor mergers, as evidenced by young globular clusters in the galaxy's halo. A much more substantial collision will happen when the Milky Way collides with our near neighbor the Andromeda Galaxy. This seems inevitable, but is not that imminent. Our best estimate is that this will take place some time in the next 3–4 billion years.

Nothing But a Gnab Gib

The science fiction writer, Douglas Adams (1952–2001), placed a restaurant called Milliways at the end of the universe. It hovered on the brink, allowing diners to view the destruction of their own galaxies and stars while enjoying beef from a cow that came to the table to advertise its best cuts and a fine selection of liqueurs from Aldebaran, Alpha Tauri. But the main characters all miss the dénouement. 'What about the End of the Universe? We'll miss the big moment,' says Arthur Dent. Two-headed Zaphod Beeblebrox answers. "I've seen it. It's rubbish. Nothing but a gnab gib…Opposite of a big bang."

Nothing but a gnab gib it might be to those who have seen it many times from their Milliways tables, but what is the ultimate fate of the universe? We noted in an earlier chapter that supernovae function well as standard candles for very distant galaxies. This was something that Fritz Zwicky pioneered. Scientists have examined how supernovae have evolved over five billion years, and they have found evidence of an acceleration in the rate of cosmic expansion. The implication of this is that, among the models for the universe's ultimate fate, relentless expansion seems the most likely.

The 2011 Nobel Prize in Physics was awarded half to Saul Perlmutter and the other half jointly to Brian P. Schmidt and Adam G. Riess "for the discovery of the accelerating expansion of the Universe through observations of distant supernovae."

Chapter 22

Nearest the Sun

Observation: Mercury and Venus
Significance: The nature of rocky planets, distance from Earth to the Sun
Science: Transits, processes on Venus, the MESSENGER mission

We return to our immediate neighborhood for our next observation, a double encounter with the two planets closest to the Sun. Venus and Mercury present a strange observing experience, and it is tempting to turn away because there is not always much to see. Don't! Look and look again, because measuring these planets and their orbits have given us some of the fundamentals of our universe.

In case you ever wondered, rendering Venus into an adjective in English is difficult. The oldest one used is Venerian, but many modern writers prefer either Venusian, a word created by Andrew Blair in his science fiction work *Annals of the Twenty-Ninth Century* (1874), or they use the noun form. We will use Venusian here. A few very old books prefer a word of Greek origin, Cytherean.

Seeing Venus

"The fairest star who stands in the sky," so the ancient Greek poet Homer wrote of Venus in the *Iliad,* for at its brightest Venus outshines every other celestial object but the Moon and Sun. Blazing at magnitude −4.6, the planet draws the eye towards it in the evening or morning sky. Since it is, like Mercury, an inferior planet – implying no value judgement, inferior meaning that it lies within Earth's orbit – it never wanders too far from the Sun, but unlike Mercury it is both brighter and more persistent.

M. Marett-Crosby, *Twenty-Five Astronomical Observations That Changed the World: And How To Make Them Yourself*, The Patrick Moore Practical Astronomy Series, DOI 10.1007/978-1-4614-6800-4_22, © Springer Science+Business Media New York 2013

To observe Venus, there is first of all a familiar caution. Do not look straight at the Sun or point a telescope anywhere near it. Venus always appears fairly close to the horizon at dusk and in the early morning. Consult a magazine or online resource to establish if Venus is appearing and then you really can't miss it, especially if you pick a time when Venus is at maximum elongation.

Let's remind ourselves what this means. Elongations occur when the orbital path of either Mercury or Venus forms a tangent furthest from the Sun as viewed from Earth. There are two kinds of elongation: eastern elongation occurs when the planet is in the evening sky, while western elongation brings the planet into the early morning sky.

So striking are these two occurrences of Venus, at morning and at evening, that they were taken to be two quite different stars. Homer distinguished between Hesperos, the evening star, and Phosophoros ('light-bringer') or Heosphoros ('dawn-bringer') in the morning. It is thought that the ancient Greek poet Ibycus of Rhegium was the first to talk of them as one and the same.

Fig. 22.1 Venus among the Pleiades, captured on 31 March 2012 (Credit: Jimmy Westlake)

Naked-eye observing of Venus is straightforward and enjoyable. It is a great feature of the evening sky to point out to others. At time of this writing, Venus is wandering through Capricorn and Aquarius.

Planetary conjunctions involving Venus often make spectacular observing targets, especially when Venus is close to Jupiter. Venus can also find itself apparently along-side other, dimmer objects, and a conjunction between Venus and Neptune is a good way of identifying the distant ice giant, or it can be positioned among the stars, as in the above image showing the planet seemingly beside the Pleiades cluster.

Through a small telescope, Venus reveals phases. These are caused, like those of the Moon, by variations in lighting. They have been recorded ever since Galileo, whose drawings you will be able to match to your observations.

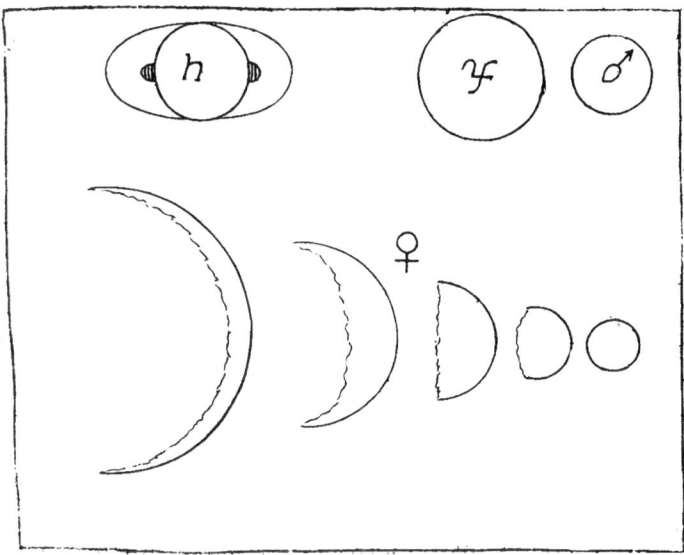

Fig. 22.2 The phases of Venus as sketched by Galileo in the *Starry Messenger* of 1610 (http://www.springerimages.com/Images/Physics/1-10.1007_978-1-4419-6803-6_1-11) (image from the Galileo project at http://www.astronomy2009.org/static/archives/images/screen/galileo_12.jpg)

The crescent Venus is especially lovely to observe, but it is impossible to see Venus in all phases of illumination. When Venus is at full, it is directly behind the Sun.

The planet is a compelling object, inspiring awe and fear in human history. In the Dresden Codex, a Venus Table created by Mayan astronomers predicts appearances of the planet as a tool for warning the community against the planet's baleful influence. This is from the Anales de Cuauhtitlan, a sixteenth-century compilation of earlier traditions:

> It was after eight days that the morning star came out, which they said was Quetzalcoatl.... When he goes forth, they know on what day sign he casts light on certain people, venting his anger against them, shooting them with darts.

Deprived of this dread, there was still the planet's beauty to contend with. It is recorded that the Emperor Napoleon was aggrieved at a crowd paying more attention to Venus than to himself.

Can we see any features when we look at the planet? Observations of marks have been made since telescopes were first trained upon the planet, and some of these marks attracted names, and rather exciting ones – the Royal Sea of King John, the Sea of Prince Constantine, the Strait of Vasco di Gama.

It was the great Russian astronomer Mikhail Vasilyevich Lomonosov (1711–1765) who cast doubt upon these observations when he suggested in 1761 that what we were seeing was the thick atmosphere of the planet and not the surface at all. The evidence supporting Lomonosov's view became too great to ignore. "As to the mountains in Venus," William Hershel wrote some years later, "I may venture to say that no eye will ever get a sight of them." He was right. No telescope can pierce Venus' clouds, and modern observations of the surface are achieved by radar.

It is possible to observe and record transient patterns in the Venusian atmosphere using UV, blue and violet filters and clever imaging software. Beware, however, of thinking you have spotted outbursts of color or wisps in the atmosphere. These are ghost effects that can appear on some telescope lenses caused by the planet's brightness.

One strange observing experience is Venus' ashen light, the faint visibility of the unilluminated part of the disc comparable to Earthshine as seen on the Moon. Sir Patrick Moore writes that "On occasions, I have seen the light so plainly that I cannot dismiss it as illusory." As we will discover at the end of this chapter, explaining this observation has led to at least one remarkable account of the planet's inner secrets.

Transiting the Sun

A very particular form of conjunction, only possible for Venus and Mercury, is when the planets cross the face of the Sun. This image, taken by NASA's Transition Region and Coronal Explorer (TRACE) spacecraft, shows the transit of Venus across the Sun on June 8, 2004, using a Hydrogen Alpha filter.

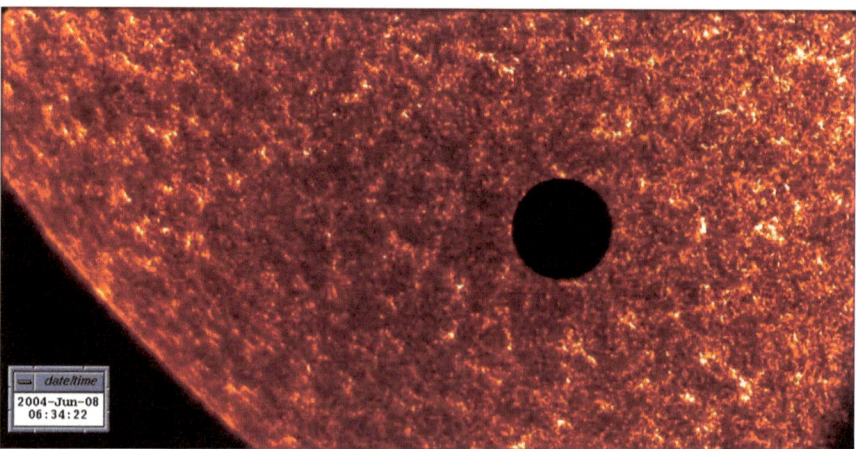

Fig. 22.3 TRACE image of Venus, 8 June 2004 Credit: NASA

As seen from Earth, these events are very rare. They can only take place when Venus or Mercury lies between Earth and the Sun and its orbit crosses the ecliptic. This event falls within a few days of either June 8 or December 9, but they are far from annual events. Kepler predicted that it would happen in 1631 but did not live long enough to see it. There were transits of Venus in 1638 and 1761 (when Lomonosov discovered the Venusian atmosphere), then 1769 and 1882, when William Harkness (1837–1903) of the U. S. Naval Observatory wrote: "When the last transit season occurred the intellectual world was awakening from the slumber of ages, and that wondrous scientific activity which has led to our present advanced knowledge was just beginning. What will be the state of science when the next transit season arrives, God only knows."

That next transits took place in 2004 and in June 2012. There is now a long gap until 2117.

Venus transits have always been important scientific events, none more so than that of 1769. Like a lot of good science, observing the transit proved an expensive business. The Royal Society in London had to petition King George III for the money needed, claiming not only that "[it would] contribute greatly to the improvement of astronomy on which Navigation so much depends" but also, and perhaps more significantly, that "the French, Spaniards, Danes and Swedes are making the proper dispositions…[and] it would cast dishonor on the British people should they neglect to have the correct observations made of this important phenomenon." The king coughed up some cash, and in August 1768 the Royal Navy ship *Endeavour* set off to Tahiti with James Cook in command. It was on this voyage that Cook mapped the eastern coast of Australia and landed at Botany Bay.

The naturalist, polymath and scientific patron Joseph Banks (1743–1820) joined Cook on his mission. In his diary he recalls how he sought to explain to the people of Tahiti what it was that the strangers were doing there: "We showed them the planet upon the Sun and make them understand that we came on purpose to see it."

The data returned by Banks and other observers enabled a fundamental unit of measurement to be determined. Using a method proposed by Scottish mathematician James Gregory (1638–1675) and revised by Edmund Halley, Thomas Hornsby (1733–1810), who himself observed the 1769 transit from the Tower of the Five Orders above Oxford's Bodleian library, was able to estimate solar parallax and so measure the distance between Earth and Sun. This distance, the astronomical unit, is now the standard yardstick for the Solar System. As Hornsby wrote in 1771 to the Royal Society.

> From the observations made in distant parts by astronomers of different nations, and especially from those made under the patronage and direction of this Society, the learned of the present time may congratulate themselves on obtaining an accurate determination of the Sun's distance as perhaps the nature of the subject will permit.

This is how Benjamin Martin (1704–1782) expressed the importance of a Venus transit:

A Transit of the Planet Venus over the Sun's Disk is not only the most rare but also the most curious Phaenomenon of the Heavens. And since the most Noble Problem in Nature, viz: the Dimension of the Solar System, is solvable only by it, & thereby renders it of the Greatest Consequence to Mankind, I concluded a proper Representation of all the Transits that happen in 1,000 Years would prove no unacceptable Present to the Public.

Hornsby's figure for the distance from Earth to Sun 93,726,900 English miles. Using Kepler's laws, Hornsby then gave distances to every known planet from Mercury to Saturn. The AU is today measured as 92,955,807.273 miles or $1.4595978707 \times 10^{11}$ m, with an uncertainty of just 3 m.

Getting Below the Clouds

Venus occupies a strange position in Solar System science. It suffers from being not-Mars. A simple comparison reveals the point. Since 1990, there have been 17 dedicated Mars missions launched, though not all of them made it, while in the same period there have been 3 Venusian probes, NASA's Magellan, the European Space Agency's Venus Express and the Japanese Space Agency's Akatsuki mission, a five-camera spacecraft directed at studying the planet's cloudtops, which at time of this writing has missed its first encounter with Venus but may achieve orbit in a few years' time.

Yet Venus matters. We have already seen how the transits of the planet in front of the Sun enabled a very significant breakthrough in the measurement of space. Venus also has the potential to teach us about ourselves, for there are striking similarities between Venus and Earth. At 12,103.6 km, its diameter is only slightly smaller than that of Earth, and both its mass and density are sufficiently similar to indicate that the planets formed in comparable ways. Venus is much more alike to Earth than any other planet we have yet found.

But piercing its clouds is difficult. The U. S. *Mariner 2* spacecraft was the first to make a successful encounter with another planet when it met with Venus in December 1962. It established that the surface temperatures of the planet were vastly inhospitable. Scientists now give the average temperature on the surface at a blistering 464° C. This puts into perspective the huge achievement of the Russian Venera missions, 16 in all, which accomplished a long list of firsts between 1961 and 1984 – first probe to enter another planet's atmosphere, first message from another planet's surface, first image from another planet. *Venera 7* survived for 23 min on the surface in December 1970, and although there were problems with the lens caps on the cameras, later Venera spacecraft were able to record and transmit pictures of Venus before they melted in the heat. The most recent landing on Venus was achieved by a probe sent from the *Vega 2* spacecraft en route to Halley's

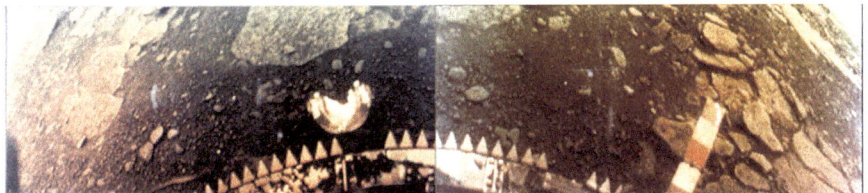

Fig. 22.4 The surface of Venus and the rim of the Venera spacecraft Credit: Soviet planetary exploration program, NSSDC

Comet. It touched down on June 15, 1985, and was able to survive for 56 min beneath an atmospheric pressure 92 times that of sea level on Earth.

Shown here is what the surface of Venus looks like, as imaged by *Venera 13.*

The study of this world is led by radar mapping, beginning with the U. S. Magellan mission in active orbit around the planet from 1990 to 1994. Magellan demonstrated the youth of the Venusian surface. In contrast to the Moon or Mars, Venus is dominated by young terrain, resurfaced around 500 Ma ago. This process was driven by widespread volcanic eruptions, and Magellan images reveal thousands of volcanoes across the planet of which the greatest, Maat Mons, is about 400 km wide and 4.5 km above the planet's mean elevation. Magellan radar images show overlapping lava flows spreading down the volcano's flanks.

Almost all the features on Venus have female names. But alongside the Australian poet Mary Gilmore and the Slavic forest witch Baba-Jaga stands one man, James Clerk Maxwell, whose Maxwell Montes (Mountains) are in Venus' Ishtar Regio. Maxwell is a hugely important figure in the formulation of theories of electromagnetism and predicted the existence of radio waves. The Working Group for Planetary System Nomenclature accepted that because our knowledge of Venus so depended on radio it was appropriate to "assign the names of deceased radio, radar and space scientists to topographic features," and gave Maxwell his place in this woman's world.

Other features confirm the power of volcanism on the planet. Radiating fault patterns called coronae, pancake domes and strange spider-like features called arachnoids are all caused probably by volcanic activity and the sticky lava emerging from the interior. Most of these have no parallels on the other terrestrial planets of our Solar System.

Is there still active volcanism on Venus? The ESA Venus Express mission in orbit around Venus at time of writing has collected evidence of at least nine current 'hotspots,' while Magellan data also suggests that some geologically recent activity has taken place near Maat Mons. Nothing rules out the possibility of volcanism being active on Venus today.

Observations of Venus reveal that the planet has a thick, enveloping atmosphere. Heat and smog drive the Venusian climate, and the clouds of Venus are mainly composed of sulphuric acid, with solid sulphur and ferric chloride giving the surface its sickly hue. Balloons dispatched into the Venusian clouds by the Vega missions measured violent windspeeds stirring an atmosphere made up of carbon dioxide with sulphuric acid and smaller concentrations of hydrochloric and hydrofluoric acid. Evidence has also mounted that Venus exhibits lightning in the clouds.

We noted earlier the similarities between Venus and Earth. Perhaps, though, it is the differences that are really more significant. Venus is 25 % closer to the Sun than is Earth, and its searing temperatures mean that there is no surface water. The carbon dioxide rich atmosphere shows a catastrophic greenhouse effect, trapping sunlight in the atmospheric vice so that the heat cannot escape.

Venus also rotates in the opposite direction to Earth and most other Solar System bodies. It does so very slowly. A Venusian day is longer than its year. The cause of this orientation is not certain, and may be the effect of an impact, although some studies suggest that friction between the core and mantle or the effects of atmospheric drag could have given the planet its unique spin.

Cloud-Life?

Venus is a radically inhospitable place. But the possibility of life on Venus was taken seriously enough during the glory days of space exploration for a group of four scientists – Wolf Vishniac, Carl Sagan, Kimball Atwood and Harold Morowitz – to form a society dedicated to examining the possibilities. "We paused to order dinner," Harold Morowitz recalls, "decided on another round of cocktails and returned to the naming problem." With the drinks came new ideas and eventually the Society of Venereal Biologists was born.

However, it never met again. "I have never put it on my resume," writes Morowitz, "fearing that promotion and tenure committees would not understand."

Yet there is a possibility of extremophile life in the clouds. The temperatures and pressures in the clouds at an altitude of between 45 and 70 km lie within the range of some extremophile bacteria known on Earth; there are tiny quantities of water there as well. If there is also lightning activity, then there should be much more carbon monoxide than we have found, suggesting, possibly, that something in the clouds is surviving off carbon monoxide and water and producing carbon dioxide and hydrogen as waste products. It is a fascinating thought and would be a testament to the power of life.

Observing Mercury

Let us now turn to the second of our two observations and seek out the smallest of the eight planets of our Solar System. Mercury is the most difficult of the classical planets to observe.

Why is it so hard? First because Mercury is the innermost planet of the Solar System and never strays far from the Sun. Whenever observing Mercury, it is essential to bear in mind the warning that applies to all solar viewing, and never to look at or near the Sun. This does not apply only to amateur observers. Hubble and other space telescopes cannot observe Mercury for fear of the damage that exposure to sunlight might cause.

Secondly, Mercury is always a near-horizon object and so has to be observed at low elevation. This is not something that large telescopes are designed to achieve. Professional CCD images of Mercury have to be assembled using millisecond exposures. The results show nothing like the clarity achieved with other celestial objects.

But it's not impossible to hunt down Mercury and enjoy what you find. Online resources and astronomical magazines publish guides to spotting the planet at those times of the year when it moves away from the Sun. It is always either a twilight or an early morning observation, and identifying Mercury demands a good idea of the pattern of background stars. At superior conjunction it is altogether concealed by the Sun, but when it reaches at its greatest elongation east of the Sun, it shines brightly enough to be seen by the naked eye, at a magnitude of around −0.7.

Is there anything to see on Mercury's surface? In 1870, two British astronomers, Warren De La Rue and William Higgins, wrote that they had identified "markings, like the lunar craters, seen as through a veil of mist," and 12 years later, the Italian Giovanni Schiaparelli (1835–1910) seemed to confirm this during a prolonged and careful study of the planet. Schiaparelli made use of the Brera telescope in Milan to do what we must never do – he isolated Mercury in the daytime sky, chasing it to within 4° of the Sun. Over many months, he became convinced that he had seen definite markings on the surface, identifying a shape of dark patches that looked a lot like the number 5. The markings were consistent, suggesting to him that Mercury was, like the Moon, locked into an orbit with one face permanently towards the Sun. Late in 1882 he wrote to a friend and described what he had seen. Somewhat unusually, Schiaparelli chose to use not only the Latin language to do this but also a poetic form called hexameters, here translated:

Mercury on its axis turns like the Moon:
One side has lasting day, the other night:
One side in everlasting fire doth swoon.
While th'other hides forever from the light.'

Schiaparelli continued to observe Mercury for years afterwards. He explained any variations in what he saw with reference to the planet's atmosphere, remaining convinced that he had seen and could still see his number 5. He built up an archive of many drawings, and in 1889 he published a map of the daytime side of the planet.

Because of the difficulties of observing it at all, Schiaparelli's conclusions regarding Mercury remained until the Mariner flybys and the advent of radio astronomy. We can now say that Schiaparelli got Mercury pretty wrong. It is also worth noting, by way of a warning, that late in his life Schiaparelli suffered from blindness, which might well have been caused by his Mercury observations.

It is spacecraft that have transformed, and are still transforming, our understanding of the planet. *Mariner 10* was able to image a part of the planet's surface during flybys in March and September 1974 and then March 1975, when it passed over the planet's north pole at an altitude of about 200 miles. It took some historic

Fig. 22.5 Messenger images the whole of Mercury (Credit: NASA/Johns Hopkins University Applied Physics Laboratory/Carnegie Institution of Washington)

photographs of the 45 % or so of the planet that it could see. *Mariner 10* is still orbiting the Sun, although its electronic systems have long been eviscerated by solar radiation.

However a much fuller encounter, one that has not ended yet, has been achieved by the MESSENGER spacecraft.

It took 2,421 days in space and 4.9 billion miles on not much more than 1,300 lb of fuel to get the MESSENGER spacecraft into a stable orbit around the little planet. Its speed exceeded of 140,000 miles per hour. In the planet's scorching environment, the spacecraft is defended by two solar panels and a sunshade. All of this has been accomplished by a machine whose clever bits weigh about the same as a largish SUV. On March 18, 2011, MESSENGER made its encounter with Mercury.

Mercury is not easy to reach. The planet's elliptical orbit confounded Newtonian physics, with its perihelion (the point of closest approach to the Sun) moving in a manner that did not fit the numbers. All sorts of theories were advanced through the nineteenth century to explain this anomaly. At one stage, observers believed that they had spotted another planet, christened Vulcan, within Mercury's orbit, proposing that perturbations caused by this new member of the Solar System were causing Mercury to misbehave.

In 1916, Albert Einstein solved the problem without any need for Vulcan. Mercury became a key observational proof of his General Theory Of Relativity, demonstrating that the planet's anomaly was just as his theory would expect. He showed that Mercury was moving in curved space, caused by the proximity of the huge mass of the Sun.

What is MESSENGER seeking to discover? First, it aims to explain the fact that Mercury is the densest of all the terrestrial planets. Measurements indicate that some 60 % of its total mass is iron core. This challenges our models of planetary formation. Are we right in what we measured? How can this have come about?

The second question is broader. It's clear from Mercury's surface that its past is not unlike that of the Moon. The terrain is ancient and has been bombarded by crater-forming meteorites. But there is also evidence of volcanic activity, a surprise for such a small world. MESSENGER will reconstruct the geological past of the planet to identify the processes that have formed the pitted and ridged surface of today.

Even the great speculators doubted that this planet could harbor life. Here, however, is Bernard de Fontenelle in his *Conversations on the Plurality of Worlds* of 1686 on what Mercury's inhabitants might be like: "They are so full of Fire that they are absolutely mad; I fancy they have no memory at all…that they make no reflections, and what they do is by sudden Starts and perfectly haphazard; in short, Mercury is the bedlam of the Universe."

The third question is allied to the first. *Mariner 10* demonstrated that Mercury has a magnetic field similar to that of Earth. This implies a semi-fluid core. MESSENGER will renew and extend the Mariner observations and use this information to explore the structure of the core and the differentiation within the planet's interior.

A fourth question concerns the material observed at Mercury's poles. Some craters are permanently in shadow, and studies suggest that they may harbor reserves of water ice. MESSENGER's cameras and other instruments will seek to ascertain if this is the case.

Finally, MESSENGER is going to measure the planet's exosphere – the atoms around the planet that are too thin to form an atmosphere and leak into deep space. Hydrogen, helium, oxygen, sodium, potassium, calcium and now also magnesium have been identified. They are coming from somewhere, presumably from within Mercury. MESSENGER seeks to explain this unstable and tenuous environment.

At time of this writing, MESSENGER has not yet finished its work. But its preliminary results propose that Mercury is a more dynamic planet than was ever previously thought. Magnometer measurements have already established that its global magnetic field is quite different from those observed elsewhere, and the surface of the planet is much less Moon-like than appearances suggest, with high amounts of sulfur and potassium detected. There is evidence for some sort of ongoing internal activity, and distinct material has been excavated from within by relatively recent impacts.

Other bright deposits seem to indicate the presence of water ice. At the 2012 Lunar and Planetary Science Conference evidence was presented that in the coldest areas of the planet's "cold pole" of 90° East, nearly all craters have radar-bright features "consistent with the water-ice hypothesis."

This is all work in progress. Much of it will be refined as new data appears. Mercury was once thought boring, but no longer. It is now one of the fastest-changing topics in space science.

Party Lights

We will end this double observation by returning to Venus. If you spend any time watching the strange patterns of its atmosphere, you will acquire a fondness for one of the planet's great admirers, Franz von Paula Gruithuisen. The son of a falconer, Gruithusien (1774–1852) was a medical scientist of note, developing both theories and practical instruments for relieving the dreadful pain caused by bladder stones.

His private fascination, though, was for astronomy. He wanted to convert the tower of Munich prison into an observatory, and when he did gain some land to build on, he used his wife's dowry to pay for his equipment. His first interest was the Moon. He is remembered in the Gruithuisen Crater and especially the Gruithuisen Domes, interesting volcanic features in the north of the Oceanus Procellarum. But he observed Venus as well and developed some interesting views on what he saw. He was especially attracted by the ashen light phenomenon, that Earthshine-like observation we noted earlier. This is how Gruithisien explained it: "We assume that some Venus Alexander or Napoleon then attained universal power…. The observed appearance is evidently the result of a general festival illumination."

So it is light from a party? Gruithuisen later changed his mind about this, perhaps worried that it seemed an improbable idea. Except his new explanation was only a little less extraordinary. He decided that the light of the planet was caused by Venusians burning jungle to produce new farmland for their expanding population.

Who says that scientists can't also be imaginative?

Chapter 23

Round and Round

Observation: International Space Station, Iridium flares
Significance: Satellites, the continued human presence in space
Science: Construction of the International Space Station, extremophile life in space

The human presence in space glimmers down at those of us who remain Earth-bound. Satellites are fleeting presences in the sky, but their effects lie all around us. It may seem strange to include a chapter on a human artifact among these 25 observations that have changed history, but few technologies have affected the modern world more than satellites. Be it where we are, who we want to talk to, what the weather will be tomorrow or how our nation's enemies are (mis)behaving, these orbiting pieces of metal define the boundaries of knowledge. The observation of satellites is straightforward and rewarding. The history of satellites and the science of the International Space Station (ISS) will determine the form and focus of humankind's future among the stars.

Stories and Scribbles

"It is the story of a great idea, a great dream, if you wish, which probably began many centuries ago…It has been dreamt again and again since…that we possibly could, and if so should, break away from our planet and go exploring to others."

So wrote the German-American writer and scientist Willy Ley (1906–1969), capturing in words the allure of rocketry, the possibility that humanity might find a means to break through the grasp of gravity and reach deep space. Because we know it has been done, it can seem obvious to us that it always would be done.

Faced with the enormous practical challenges, many, and especially those working in what was then the Soviet Union, drew inspiration from the great dreamer-scientist of rocketry, Konstantin Tsiolkovsky (1857–1935). Tsiolkovsky was a theoretician, a model-maker and a scribbler. He drew when he was not writing, and he used both words and images to test what might work.

His most ambitious novel was *Beyond the Planet Earth*, published in 1923. In it, six scientists find themselves in a Himalayan castle with unlimited engineering resources. They are drawn from all over scientific history: Laplace from France, Helmholtz from Germany, Ivanov from Russia, Newton from Great Britain, Galileo from Italy and Franklin from the United States. The Russian builds a rocket and then explains his design. This is a novel full of lectures and mathematics, where characters say things like "Let me demonstrate by writing out a table, and the corresponding speeds and distances." Then they do just that. His rocket is powered by a blast mechanism shielded by a chamber of vaporizing liquid, with zero-gravity liquid for the astronauts to move in. Tsiolkovsky conceives of an automatic pilot and an on-board telescope, filters to clean the interior atmosphere, even a greenhouse to grow supplies.

The scientists set off. Once beyond the atmosphere, they undertake a spacewalk, and people on Earth see the craft as it orbits. Messages are sent to and from the ground by using Morse code. They make it to the Moon where "They looked upon the strange beauty that no human being had seen before." Being good scientists, they then take lunar samples – no playing golf for them – and then extend their mission to Venus and Mars before returning to Earth, promptly catching the flu. No matter, for: "The Earth is still ours; and if we can't bear to be separated from her, she will always open our arms to take us back again."

Tsiolkovsky's work is utopian. He dreams of a communion of humanity forged by scientific endeavor and the colonization of space. The more important point, though, was that he used his writing as an exploration of the possible. "First the idea must be conceived," he wrote in 1926, "almost like a fantasy, then with scientific work and calculation the idea is crowned."

Other great writers of science fiction have worked in the same way. From Jules Verne to Arthur C. Clarke, the empty page has been a place to make real the impossible and to test science. Tsiolkovsky was the first to conceive of a rocket powered by liquid oxygen and hydrogen propellants. His equation for jet propulsion is invoked by modern aircraft design. He is an example, perhaps the best one, of how great science can often start in the imagination.

The journey from science fiction to satellite launch was arduous. But it was achieved 100 years after Tsiolkovsky's birth when *Prosteyshiy Sputnik,* literally 'Simplest Satellite,' was launched in the nose of an R7 ICBM on October 4, 1957. "Mankind will not remain on Earth forever," Tsiolkovsky had predicted. He was proved right that day.

We are not going to delve deeply here into the history of spaceflight. But it is worth noting something of what satellites have done to change the way we live. In at least four ways, they have transformed the developed world: in navigation, in communications, in weather and in war. The last of these we can perhaps put aside, but the earlier three are with us always. As Arthur C. Clarke, another great scientist-dreamer, wrote in 1945: "A true broadcast service giving constant field strength at all times over the whole globe would be invaluable, not to say indispensible, in a world society."

It is interesting to hear again, with Clarke as with Tsiolkovsky, the wider aspiration. They hoped that satellites would change technology – they have – and so change humanity as well. Perhaps that part of the dream still seems far away. But few things have come closer to realizing the aspiration of international co-operation than the ISS. After observing it, we will explore its history and its impact.

Observing Satellites

Observing manmade satellites is all about timing. In fact, these are not good targets for a telescope. Binoculars are more mobile and the ISS is, after all, just grazing the top of the ground compared to the stars, nebulae and planets we have observed. It orbits at an average height of 240 miles.

This does not make it a straightforward target. The ISS represents a challenging observation because it requires us to be precise as to time and location. Satellites are visible when they are lit by the Sun, but the observer is in darkness. This combination usually spans an hour or so after nightfall and another hour before dawn. We are looking for reflected sunlight – getting the time right is essential.

Here are some of the many useful websites for astronomers who want to observe the ISS:

- http://spaceflight.nasa.gov/realdata/tracking/index.html provides NASA data for the ISS, including its altitude and orbital path map.
- The Skywatch tool is at http://spaceflight.nasa.gov/realdata/sightings/index.html. There are also NASA apps for tablets.
- www.heavens-above.com for daily sky charts showing the ISS and much else.

These are good observations with which to begin, or conceivably end, a star party. There have been crews on ISS since November 2000, and the sight of humanity's reaching into space can be an inspiring start to an evening among the stars.

All satellites have precise orbits, and an orbital path map for ISS is readily available online. NASA also provides without charge a downloadable application called Skywatch. There are similar applications for iPads and other devices, which will provide you with a table showing the ISS pass times above your location. A similar provision is made by the Heavens-Above website, which also provides a wealth of detail about other satellites, comets and minor planets.

All of these resources give a time and then two other critical pieces of information: azimuth and elevation. An object's azimuth is its distance in angular degrees measured in a clockwise direction from north. Observers in northern skies can use Polaris as a marker; elevation or altitude is its angular distance, also in degrees, above the horizon.

There are some weeks when the ISS is high in the sky. These are the nights, or sometimes early mornings, to look out for because you will probably be able to dispense with worrying about azimuth and elevation and simply look in the vaguely right direction for a 'moving star.' Under good conditions, and especially when the ISS is right overhead, it is possible to make out structural details, especially its solar panel 'wings.'

Observing smaller satellites uses the same techniques as above. The Heavens-Above website is a good guide.

Fig. 23.1   An Iridium satellite flashes among the stars of Cassiopeia (Credit: Tom Martinez)

One particular target that can make for fine photographs is an Iridium flare. The one captured above shows how bright they can be compared to the w-asterism of Cassiopeia above the trees. The Andromeda Galaxy is also visible as a smudge below and to the left of the flare.

What are Iridium flares? They are not flares so much as glints, sunshine caught on the highly reflective aluminum antennae of an Iridium communications satellite. These low-orbit craft form a network used by mobile phones and other devices, and it is a peculiarity of their design that they exhibit sudden flare-like flashes from time to time. There are predictions of Iridium flares offered on the Heavens-Above and other websites based upon precise calculations of the angles made by the antennae and the Sun. It's essential, if you want to catch one, to be looking in exactly the right place at precisely the right time, as the flares are spectacular but very brief. It is a good exercise in measurement to take the predicted time and angles and set yourself up for the event, even if tiny variations in orbit can lead to disappointment.

The Asteroid That Wasn't

In June 2010, the Catalina sky survey added a new member to our Solar System. It wasn't much – an asteroid just a few meters across, not especially bright but moving quickly along its orbit.

Or was it? The orbital numbers looked strange, and the way it intercepted Earth's passage around the Sun suggested that they should have impacted within the past few thousand years. How had it survived? It proved tough to observe at magnitude +18. Spectroscopic tools had to be employed before it became clear that asteroid 2010KQ was not an asteroid at all.

It was in fact residue from human activity in space. Now merely object RK252A5, those wanting to establish what this not-asteroid once had been had to work back through its orbital history. It became clear that it had left Earth's imme-diate environment in April 1975, and it's likely that the once 2010KQ is in fact the final rocket stage of one of the Soviet Luna probes from 1959, 1963 or 1974 that had entered a highly elliptical orbit around Earth and was then perturbed into the orbit of the Sun. It was a relic of the days of lunar exploration carrying a little bit of our history into space.

We Ain't Gonna Do It with the Tools We've Got

The attempt to create a long-term human presence in space has proved much more difficult than leaving our junk behind us. It has taken some remarkable scientists and astronauts to achieve this. And what has emerged is quite unlike what was once conceived.

Its beginnings lie with the Soviet *Salyut 1,* launched on April 19, 1971, between Apollo missions 14 and 15. The first crew sent up to occupy it were not able to access the docking collar, but in June three astronauts, Georgi Dobroveolsky, Viktor Patsayev and Vladislav Volkov, did enter from the *Soyuz 11* spacecraft, and they made something like a home there. Their compartments were small, tiny by comparison with what came later, but Salyut included a space observatory, Orion 1. Viktor Patsayev was the first man to operate a telescope outside Earth's atmosphere, obtaining spectrographs of the stars Beta Centauri and Vega. But the astronauts were guinea pigs, essentially, for no one knew if anyone could survive for any length of time in space. They lasted for 23 days until a small fire on board prompted an early return. On June 29, they left Salyut for Earth. They were found dead in the descent module when it landed.

This early tragedy bears witness, and stark witness, to an important fact. However routine it has become, long-term missions into low orbit have been and still are dangerous.

Salyut 1 was decommissioned. The Soviet space station mission continued, with several further Salyut missions of which the longest and most successful was *Salyut 6,* which was used from 1977 to 1982 by a succession of crews. The space station included both an internal and an external telescope, as well as cameras for Earth observation. After their earlier disaster, the Soviets established a reliable system for resupply using their Progress freighter spacecraft, which under various designs has continued to supply space stations since.

The American program began with Skylab. Its aim was to create a platform for biomedical experiments, Earth surveys and the study of the solar corona. It included a compartment some 150 times bigger than the Apollo command module. Launched in May 1973, it suffered immediate problems with its heat shield systems and required an urgent repair mission. Launched on a Saturn IB rocket from the 'milkstool' that enabled it to use the Saturn V facilities, the SL-2 team docked their Apollo module with the nascent space station and were met by temperatures of 50 ° C in the living quarters. Astronauts Pete Conrad, Paul Weitz and Joseph Kerwin had to erect a sunshade and clear away other damaged sections, leading Conrad, previously famous for his remark "Never come to the Moon without a hammer" to warn Mission Control: "We ain't gonna do it with the tools we've got."

In fact they did, and the SL-2 repair mission achieved not only the rescuing of Skylab but established what was to become an important aspect of space station design, the use of spacewalks.

Skylab was not intended to last. It was powered down after the third crew left in September 1974. Yet its achievement endured. Skylab made the vision possible, and NASA never forgot. Work on new models for a space station continued, receiving a huge boost on January 25, 1984, when President Ronald Reagan spoke of how "We can follow our dreams to distant stars, living and working in space for peaceful economic and scientific gains." This slid from vision to reality when he announced: "I am directing NASA to develop a permanently manned space station," Reagan later naming it the Freedom Space Station. The Senate, though, was not persuaded, and cuts meant that some $11.4 billion was spent without any flight-ready hardware being produced.

The Russians had much more success with their Mir station, adapted from the earlier Salyut program. It opened a new way forward for space stations in that it was assembled from separate modules rather than being launched as a single piece. Its core was lifted into orbit in 1986, but the whole was not completed until 1996.

Mir straddled huge political changes in Russia. There were debates as to whether it might be abandoned, adapted or even sold to the Americans. In July 1991, agreement was reached to share aspects of its use, and from 1995 the U. S. space shuttle was used to complete the building program. Mir also was able to attract funding from other space agencies. This merging of interests around a single orbiting platform opened the way for the creation of the International Space Station.

The ISS might even have been 'bootstrapped' onto Mir. It wasn't. Mir was mothballed in June 2000 and de-orbited the following year, but it was a Russian module, Zarya, which formed the initial piece of the ISS jigsaw. This was launched on November 19, 1998. NASA administrator Dan Goldwin commented: "We only have 44 more launches to go, and about 1,000 h of spacewalks, and countless problems, and countless issues."

This was a realistic prognosis. The whole program was afflicted by budgetary constraints, and many of the more ambitious hopes for ISS, like its predecessors, had to be abandoned. There were many problems and some near disasters, but none of them were countless. The International Space Station did grow, in large measure due to the labors of successive space shuttle missions. Zarya was mated with the American Unity module, and the first resident crew was established in October 2000. Gradually, the ISS came to assume its now familiar winged shape, incorporating additional U. S., Russian, Japanese and European units. As of August 2011, there have been 135 launches to the ISS and the total assembly including its large solar arrays spans the area of a typical American football field. "The complex now has more liveable room than a conventional five-bedroom house," NASA assures visitors to its website, including two bathrooms, a gymnasium and a 360° bay window. It is possible to get very close to their living conditions, watch them at work and read daily reports on their activities via the NASA website or iPad app.

And Now?

"The end of the beginning of humankind's emergence from its cradle." This is how David M. Harland and Robert E. Catchpole summarize the achievement of building the ISS in their *Creating the International Space Station* (Springer, 2002). If the infancy is over, where is the ISS going now?

Certainly, the ISS is capable of performing useful science. It already has done so, especially in the growing understanding of human endurance in space, both physical and psychological. It is the essential laboratory for missions to establish a permanent presence on the Moon or to reach Mars. If the ISS has felt a little like an end in itself, that is perhaps because the process of putting it together has been difficult and marred by tragedy. The technical reality has been achieved, but the vision, somehow, has yet to grow.

Will this observation change the world? Satellites certainly have. Their failure or destruction has become a doomsday scenario for our civilization. As for the ISS, it is there to be seen, and might yet open out new scientific and medical possibilities. It might also prove to have been a costly leviathan. We shall see.

Fig. 23.2 The international space station (Credit: NASA)

A Little Hope

There is some good news, though, for those who long to live in deep space. We now know of one animal that can survive there.

And this is the tardigrade, or water bear, a microscopic invertebrate distributed very widely across Earth that has the ability to survive waterless and freezing conditions. In September 2007, tardigrades were exposed to the vacuum of space in the BIOKON pack, part of the Russian FOTON-M3 mission. For 12 days these little

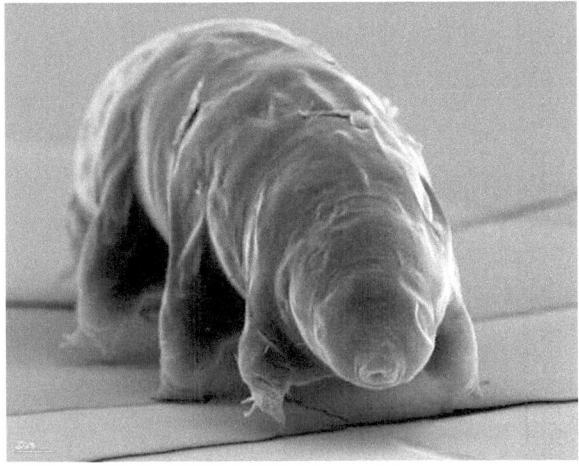

Fig. 23.3 One among 1,100 or so species of tardigrade (Credit: author's collection)

animals were studied in low-Earth orbit under the bombardment of solar and cosmic radiation, some 250 and 290 km above their normal habitats.

The results were interesting. Those tardigrades that had already entered their desiccated state survived as well in space as in laboratory conditions. They proved resistant to the demands of the environment. The hydrated ones did not do badly either, and even showed a greater tolerance to radioactivity than their dried-out colleagues. More tardigrades were sent up to the ISS as part of the BIOKIS mission in 2011.

Tardigrades may be small, but they are tough, perhaps the toughest animals on the planet. They may offer some clues as to how life on Earth might reach further into space.

Chapter 24

Finding Planets Around Other Stars

Observation: Constellation Gemini, M15
Significance: The search for other planets and life outside the Solar System
Science: Exoplanet detection methods, the Kepler mission

In 1952, the astronomer Otto Struve wrote that: "One of the burning questions of astronomy deals with the frequency of planet-like bodies in the galaxy which belong to stars other than the Sun.... But how shall we detect them?"

The lure of identifying planets outside the Solar System has drawn dreamers to dream, theorists to predict and astronomers to create means by which orbiting objects could be detected in the glare of their stars. In the late eighteenth century, Sir William Herschel thought he had found one after spotting an orbital deviation in a binary star system, and this same star, 70 Ophiuchi in the constellation Ophiuchus the Serpent-Bearer, behaved erratically enough for the American astronomer T. J. J. See (1866–1962) to declare that "the system contains a dark body." But by 1952, there were still no confirmed planets beyond the kingdom of our Sun, prompting Struve to propose a method that might yield results. These are his words from the same article: "There seems to be no compelling reason why the hypothetical stellar planets should not, in some cases, be much closer to their parent stars than is the case in the solar system.... If the mass of this planet were equal to that of Jupiter, it would cause the observed radial velocity of the parent star to oscillate... [which] might just be detectable."

Struve was right – his method works. So, too, with the telescopes at the disposal of astronomers today, as does Herschel's astrometry. In fact, the first definite exoplanet was identified in the orbit of a pulsar in 1992, while the first planet orbiting a main sequence star was 51 Pegasi, confirmed in 1995.

Thirteen years later, in 2008, we come to the remarkable image shown here.

M. Marett-Crosby, *Twenty-Five Astronomical Observations That Changed the World: And How To Make Them Yourself*, The Patrick Moore Practical Astronomy Series, DOI 10.1007/978-1-4614-6800-4_24, © Springer Science+Business Media New York 2013

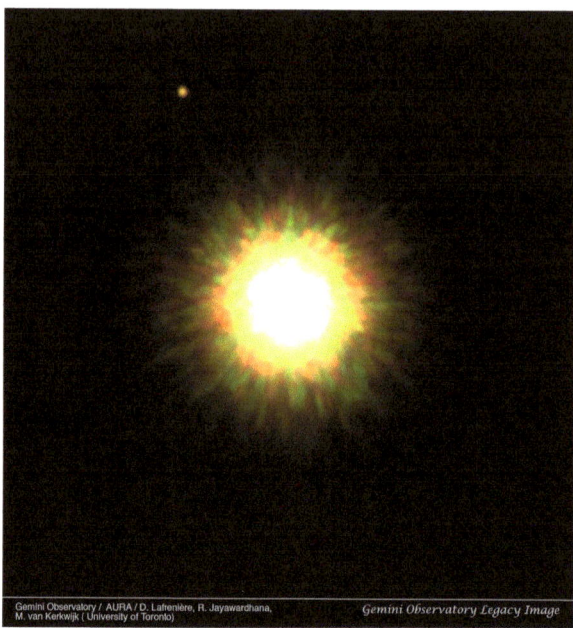

Gemini Observatory / AURA / D. Lafrenière, R. Jayawardhana, M. van Kerkwijk (University of Toronto)   *Gemini Observatory Legacy Image*

Fig. 24.1 Alpha Piscis Australis, the star Formalhaut, with a possible planet (Credit: Gemini observatory/AURA/D. Lafreniere, R. Jayawhardhana, M van Kerkwijk (University of Toronto))

Taken using the adaptive optics system and the Near-Infrared Imager (NIRI) on the 8.1-m diameter Gemini North telescope on Hawaii, it shows the star Fomalhaut, Alpha Piscis Australis, with what is thought to be a planet in its orbit.

At time of this writing, some 770 exoplanets are known to exist. Many thousands of candidate planets await confirmation. By the time you read this, these numbers will be out of date. No field of astronomical discovery is moving so fast and has such far-reaching implications as the search for planets around other stars. Our penultimate observation is changing our understanding of the Solar System and renewing the search for life.

There are four observing activities in this chapter. They will take you near exoplanets, but only near them. Observing planets around other stars is difficult. They are dim against the brightness of their suns, and they are small. Many are very far away. They are all beyond the reach of amateur telescopes. But watching the stars is an important way that exoplanets are found, and there are ways in which amateur observers can play a part even in this branch of our science.

The Twin's Friend

It is late in the life of this book to observe the constellation Gemini. Castor and Pollux are easily identified by the naked eye. Pollux shines orange as befits a K-type star, while Castor is a bright white member of the A spectral type. Unusually, the twin stick men stretching down from these two stars actually look like their constellation, their feet touching the topmost rim of Orion, the twins divided by a line between Betelgeuse and Pollux that passes through Gamma Geminorum, the star Alhena.

Through a small telescope and in reasonably good seeing, Castor quickly separates into two bright stars and a third dim star, respectively Alpha Geminorum A, B and C. The Castor system, some 45 light years from us, is made up in fact of six stars, for each of these three are themselves double stars rotating in close orbits of 9 days, 3 days and, for Alpha Geminorum C, less than the span of a single day. Neighboring Pollux (Beta Geminorum) seems lonely alongside this busy system.

In fact it isn't. In 2006, it was confirmed that Pollux hosted a single substellar companion. The planet, known as Beta Geminorum B or HD62509b, has a mass some 2.9 times that of Jupiter. It is in almost circular orbit around the star at a distance of 1:69 AU. In 2012, a study established the stellar mass of Pollux to be 1.91 times the mass of the Sun, with an error of just +− 0.09. This was the first time a giant star with a planet had been measured in this way.

It is worth taking the time to observe further in Gemini. Both Castor and Pollux have stars near the line of the ecliptic that mark their 'waists,' and that beneath Pollux, the star Delta Geminorum or Wasat (its Arabic name means "middle") is a useful guide star to a lovely deep sky object. Placing Wasat in the center of the finderscope, it is not too difficult to identify a neat triangle of stars with Wasat in one corner. The others are 56 and 63 Geminorum. Move away from Orion and have a go at splitting 63 Geminorum before continuing in the same direction to the smudged disk of NGC2392, variously known as Caldwell 39, the Clown Face and the Eskimo Nebula. At magnitude +9.2 it is not the easiest object to keep in view, but it rewards high magnification, and the central star does not blot out the blue-gray nebulosity around it. What you won't see, by the way, is the detail of the face-like features or the furry-rimmed hood that give it its names. These only appear in professional images.

It was very close to Wasat that Clive Tombaugh identified the planet, now dwarf planet, Pluto by comparing photographic plates. More permanently close to the star, but a very challenging observation, is the flat spiral galaxy NGC2357 to its northeast.

Not many constellations become oaths, so before leaving Gemini we should note that the mild exclamation 'By Jiminy' should be favored by all astronomers when such phrases are required. It is thought that 'By Jiminy' and 'Jiminy cricket' both derive from a seventeenth-century 'Oh Gemini,' perhaps an astrological imprecation or just because it was pleasing to have an expletive constellation.

Doppler Spectroscopy

Although we tend to think that a planet orbits its star, in fact they both obit around their common center of mass. To an observer, this means that the star appears to wobble backwards and forwards under the influence of the planet. So with the correct line of sight from Earth, the star appears to move towards and away from us. These variations are small, but astronomers seek out displacements in the spectral lines exhibited by the star, a Doppler effect.

This method has proved very successful. Around 90 % of all extrasolar planets thus far identified have been detected using this approach, including that around Pollux. This does not make Doppler spectroscopy, which is also called the Radial Velocity Method, easy. Our own massive gas giant Jupiter would cause the Sun to move by 13 m/s over 12 years, so detectors have to achieve extreme accuracy. Systems such as HARPS at the La Silla Observatory in Chile or HIRES at the W. M. Keck Observatory on Mauna Kea, Hawaii, are able to detect changes in velocity as small as 1 m/s.

As with all methods, Doppler spectroscopy has to find ways of excluding false results. There are other sources of spectral shifts, for example sunspots or plages (discussed in Chap. 2) and binary companions, perhaps very small white dwarfs, in orbit around the star.

An interesting consequence of Doppler surveys indicates a correlation between orbiting gas giants and parent stars exhibiting high metal content. By metal, astronomers here mean elements that come after hydrogen and helium in the Periodic Table. This may represent either a residue of the original material out of which the protoplanetary cloud formed or an effect of the inward migration of the gas giants. The former is more likely. It has been suggested that the probability of a giant planet is proportionate to the square of the number of iron atoms found in a stellar atmosphere.

Using the Square

The next observation takes us to the Square of Pegasus, an easily recognizable asterism that we have made use of before in observing the Andromeda Galaxy. The most distinctive star in the quartet is Beta Pegasi. It is an M-type red giant some 200 light years from the Sun with an unexpectedly cool surface temperature. It therefore shines more prominently in the infrared than in the visible. Alpha Pegasi, the star Markab, forms one side of the square with Scheat, and a line between the two of them leads to two stars that are important for exoplanetary science.

Beta Pegasi bears the name Scheat, meaning *shin*, ascribed to this star in error. According to the *Almagest* it belongs to the star Delta Aquarii but was applied to this star by accident.

About half way down between the stars and outside the borders of the square sits 51 Pegasi, the star that holds the distinction of being the first whose planet was ever detected in 1995. The star belongs to the G spectral type along with our Sun, and the planet has a mass about half that of Jupiter. But it orbits its star in just 4.2 Earth days.

It is argued that this planet, along with many other giants, has migrated inwards over the course of its life. This planetary migration is now an important additional element that has to be incorporated into models of our own Solar System.

Meanwhile within the square and just about opposite 51 Pegasi lies the star HR8799. It is rather unexciting in appearance through an amateur telescope, although it is an unusual young star. Unlike the majority of planet-carrying stars, it does not exhibit any metals in its upper atmosphere. We will return to HR8799 in the next section.

While in Pegasus, don't miss the opportunity to observe the Messier object M15. If you draw a straight line from Gamma to Alpha Pegasi and then carry on, you'll spot a smudgy star if the night is clear. Through binoculars it remains a single point of light, but a small telescope reveals it to be a globular cluster. It lies in the galactic halo of our Milky Way and may contain as many as 450,000 solar masses, many concentrated into a remarkably dense core. Among these stars may lie uncounted planets contributing their shards of brightness to the cluster's glow.

Precise Measurement and Direct Sight

Herschel was not wrong when he proposed that careful measurement could reveal the presence of exoplanets. Although astrometry has not yet produced certain detections yet, the Hubble Space Telescope's Fine Guidance Sensor has been used as a tool in confirming candidates identified by other means, and future space-based astrometric surveys are likely to provide the kind of detail that will reveal new exoplanets.

Neither Herschel nor Otto Struve could have conceived it possible that we would be able to image exoplanets. Yet it can be done. One example may be that remarkable Gemini image of Fomalhaut. The star certainly has a detectable debris belt, where a huge amount of activity is taking place. One recent study has estimated that the equivalent of 2,000 comets of 1 km in diameter are destroyed by interactions within the belt every day.

Observing Fomalhaut is easy if you are at the right latitude to be able to see its constellation. Fomalhaut lies in Piscis Austrinus (sometimes Australis), the Southern Fish, Lying beneath Aquarius, Fomalhaut is the brightest star in the neighborhood. Its name means "mouth of the fish."

Until very recently, some special circumstance was required to make photography like this possible. A very large planet separated by a huge distance from its star might mean that it might escape the glare of the star. But in 2010 NASA combined adaptive optics with a coronagraph, blotting out starlight so as to let the planet or, in this case, planets shine. This is HR8799, with the star smothered and the planetary system visible.

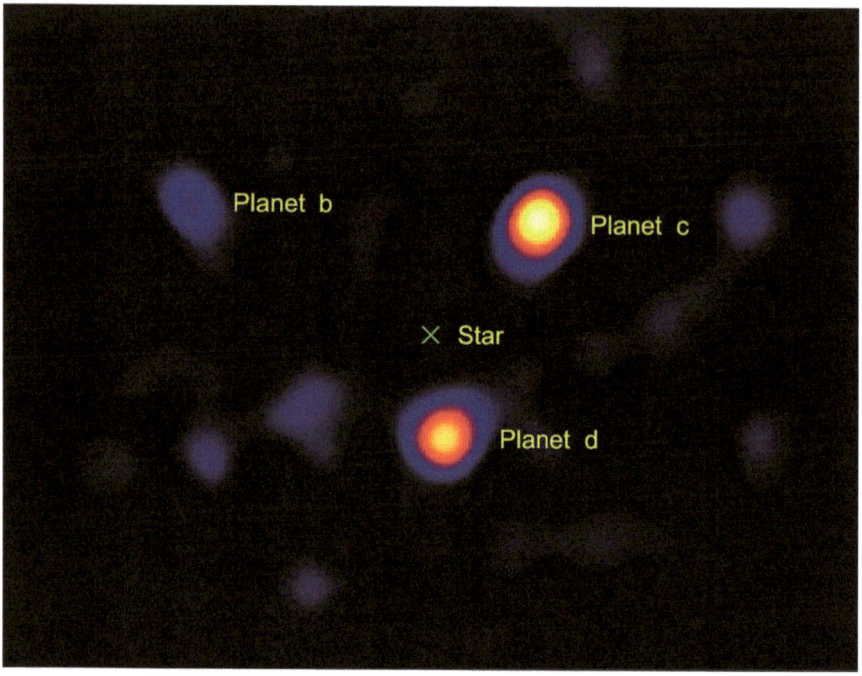

Fig. 24.2 The light from three planets orbiting a star 120 light-years away. The planets' star, called HR8799, is located at the spot marked with an "X." This picture was taken using a small, 1.5-m (4.9-ft) portion of the Palomar Observatory's Hale telescope, north of San Diego, California (Credit: NASA/JPL-Caltech/Palomar Observatory)

The achievement represented by this image is immense. It opens a new field for astrophotography as radical in its consequences as that achieved by the Drapers in their day. The image shows no star by virtue of the coronagraph, but the planets b, c and d are visible. They have long orbits, respectively 100, 200 and 400 years. A fourth planet, HR8799e, is much nearer the star and is not visible in this image. Nor is the star's debris disc.

In fact, an older direct image of this system has existed since 1998. It has recently been discovered within data acquired by the Hubble Space Telescope. A new technology for retrieving such details opens the possibility that Hubble may become an important source of exoplanet observations.

The smallest exoplanet to be observed directly thus far is circling the star Beta Pictoris. Beta Pictoris lies close to Canopus, the mightiest star of the southern sky and alpha star of Carina, the constellation of the Keel. If you can identify Canopus, the second brightest star after Sirius, then Beta Pictoris in the constellation Pictor the Easel is just above, in the rather empty area of the sky beyond crowded Carina. Like many of the stars with planets, it is not itself a fascinating target for observation. But it does deliver images like this.

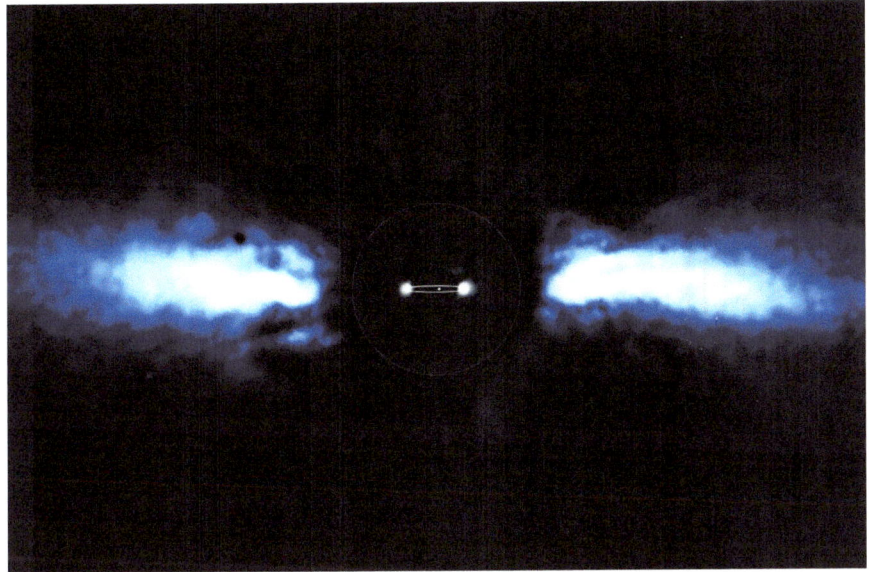

Fig. 24.3 Beta Pictoris, imaged using the NAOS-CONICA instrument (NACO) on one of the 8.2-m unit telescopes of ESO's Very Large Telescope (VLT) (Credit: ESO/A.-M. lagrange)

What are we seeing here? This European Space Agency image is a composite of pictures taken in 2003, 2008 and 2009. It shows a small light source first on one side of the star and then the other. The possible orbit of the planet is also indicated, for this is the first image of a exoplanet's orbit.

Meanwhile it is expected that the Gemini Planet Imager, deployed on the Gemini South telescope in the Chilean Andes, will begin science operations late in 2012. A number of the instruments on the James Webb Space Telescope will include coronagraphs such as that used in the HR8799 picture above. We are likely to receive more direct images of exoplanets as the technology evolves.

The Lights Go Down

The most intuitive method of identifying exoplanets comes from something familiar, the observation of an eclipse. When the Moon passes in front of the Sun, the light of the Sun is occluded. So, too, with planets passing in front of their stars. If you measure the light curve with sufficient accuracy, there is a dip each time the planet blocks the view, and a smaller dip when the planet, on the far side of the star, no longer adds its little light.

This is the transit method of detecting exoplanets. It requires an alignment between the observer and the orbital path, but if that is in place, then it can reveal a wealth of detail about the planets: orbital inclination, planetary radius and mass.

It is therefore possible to calculate the density and surface gravity and the nature of the planetary atmosphere. The French Space Agency's CoRoT mission, launched in 2006, combines the search for exoplanets with studies of stellar seismology. NASA's Kepler mission is the subject of a separate section a little later in the chapter.

The transit method is biased towards those planets with shorter orbits. They cross in front of their stars more frequently. It is again important to exclude false positives, especially grazing eclipses in binary systems, but the opportunities presented by this method are considerable, especially using surveys that can monitor large numbers of stars simultaneously over long periods. At the time of writing, some 135 planets have been identified by transit. They range in size from CoRoT36, a behemoth some 21 times the mass of Jupiter, to Keplter-11f, just 2.6 times the diameter of Earth.

How Many Exoplanets Are Out There?

This is the successor to Otto Struve's question of 1952. We know that there are planets and we have methods to detect them, but how frequently do stars have one or more planets in their orbit?

Perhaps the nearest we can get to an answer at this time comes by way of a further detection technique, gravitational microlensing. If two stars are aligned such that the light of the further shines through the light of the nearer, then the mass of a planet in the foreground star bends the light coming from the background star, and the Theory Of General Relativity predicts very precisely how that light should behave. By this means, it has been possible to identify a number of exoplanets. The principal surveys are MOA (Microlensing Observations in Astrophysics) and the charmingly initialed OGLE (Optical Gravitational Lens Experiment). Using data collected by OGLE between 2002 and 2007, a study has suggested that around 17 % of stars host a Jupiter-like planet within 0.5–10 AU, with a further 52 % having a Neptune-like planet in the same range. The survey concludes that: "This result is consistent with every star of the Milky Way hosting (on average) one planet or more in an orbital-distance range of 0.5–10 AU. Planets around stars in our galaxy thus seem to be the rule rather than the exception."

More microlensing data seems to confirm this picture of abundance. Working with the HARPS instrument at La Silla in Chile, it is estimated that some 40 % of red dwarf stars have a 'super-Earth' orbiting in the habitable zone where liquid water might exist on the surface.

Goldilocks

This evidence places a numerical value on one element in the Drake equation that we considered in an earlier chapter. It then raises Drake's next parameters – how many exoplanets are Earth-like and capable of harboring life?

We are looking for planets in the Goldilocks zone. Like the porridge in the fairy tale, they need to be neither too hot nor too cold. Distance from their star is a crucial factor in the possibility of life, especially if that life is, as we thus far believe it must be, associated in some way with the presence of water.

We are also looking for planets that are the right age. Some of the stars where exoplanets have been identified are very young, pushing to the limit our model for the formation of a solar system and the timescales involved. This is especially the case as regards the time we think it must take to form a Jupiter-like gas giant.

Hold on, though – is there life on Earth? This may seem a strange question, but we noted in an earlier chapter that the Galileo probe did not detect us when it turned its instruments on our own planet. If the acknowledged goal is to find exo-Earths and we are looking for likenesses, then a short piece of fiction written by Arthur C. Clarke in the early dawn of the Space Age becomes relevant.

There are few writers who have so successfully melded good science with fine fiction as Sir Arthur C. Clarke (1917–2008). Isaac Asimov (1920–1992) was, though, another. This is the dedication in Clarke's volume of essays *Report on Planet Three and Other Speculations*: "In accordance with the terms of the Clarke-Asimov Treaty, the second-best science writer dedicates this book to the second-best science-fiction writer."

His *Report On Planet Three* purports to be a document discovered in the post-thermonuclear wreckage of Mars, the Late Uranium age of its civilization, about 1,000 BCE in Earth dating. It is a sober assessment, using the best telescopes available at the time, of whether there is any evidence for life on Mars' near and mysterious neighbor, a sickly blue-green planet with a silver moon beside it.

Their findings are not optimistic. Two-thirds of the planet is covered with a liquid they identify as water. Reflecting their own experiences on Mars they write: "Our largest telescopes revealed intolerable temperatures on its equator; at higher latitudes, however, conditions are much less extreme, and the presence of extensive ice caps...indicates that temperatures there are often quite comfortable." The weight of Planet Three's gravity also concerns them, but it is the poisonous atmosphere that leads them to conclude that: "Looking at these well-established facts, we can now weigh the prospects for life on Earth. It must be said at once that they appear extremely poor; however, let us be open-minded and prepared to accept even the most unlikely possibilities, as long as they do not conflict with scientific laws."

Oxygen, they argue, cannot exist consistent with life. The ozone layer would block out all the UV radiation upon which Martians flourish. There would be, under Earth's oxygen mantle, something called fire. Simulated in Martian laboratories, this is devastatingly destructive. "Fire...must be quite common on Earth," their experiments suggest, "and no possible form of life could exist in its presence." Flying animals might be able to escape it and make the vast migrations necessary to avoid the awful heat. "The conception of flying animals, though a charming one, is not taken seriously by any competent biologist."

The only hope of life, then, lies in the unfamiliar habitat of water. Here the Martians are not sure. "Strange though it is for us to imagine creatures that could live in water, it would seem that the seas of the Earth may provide a less hostile environment than the land," they conclude. Rockets, when powerful enough to cross the cosmic gulf, will be able to assess the oceans for evidence. The chances of finding anything are slight, and so: "We must resign ourselves to the idea that we are the only rational beings in the solar system."

The point is neatly made. They know much more of Planet Three than we do of any exoplanet. Their conclusions, drawn out of their assumptions as to how life must work, lead to elegant but wrong conclusions. Ours might be equally elegant, and just as mistaken.

The Kepler Mission

NASA's Kepler mission has already identified a Goldilocks planet. By the time you read this, there may be more. Launched in March 2009, the spacecraft's objective is "to find terrestrial planets...especially those in the habitable zone of their stars where liquid water and possibly life might exist." Of the 61 planets confirmed at time of writing (a further 2,300 are currently candidate planets awaiting further investigation), a group have been identified that fulfill the basic parameters. They are rocky planets where liquid water might be present. The first of these, Kepler-22b, was a discovery when "Fortune smiled upon us," according to Kepler principal investigator William Borucki. Its first transit was spotted a mere 3 days after science operations began.

Kepler has been looking at a small patch of sky between Deneb and Albireo in Cygnus. It produces data by way of the transit method, gathering information on more than 150,000 stars. As of now, the Kepler mission is in safe mode and may be at its end. But it has been a transformative mission.

This is the light curve for another rocky planet, Kepler-10c, which has a radius of 2.2 times that of Earth and orbits the star every 45 days. It is almost certainly too hot for life. Kepler has identified new classes of planet, and in one case has seemingly confirmed one of the great moments of science fiction, when Luke Skywalker looks out at the double sunset over his home planet of Tatooine. Kepler-34 b and Kepler-35 b are both Saturn-like and in orbit around a pair of Sun-like stars. These stars are in motion, eclipsing one another every 28 days.

It seems impossible given the distances and accuracy required, but amateur astronomers can play a role in this work. The first is by way of light curves being made available through the Kepler mission and the planet-hunters website. The human ability for pattern recognition can spot evidence that computers miss.

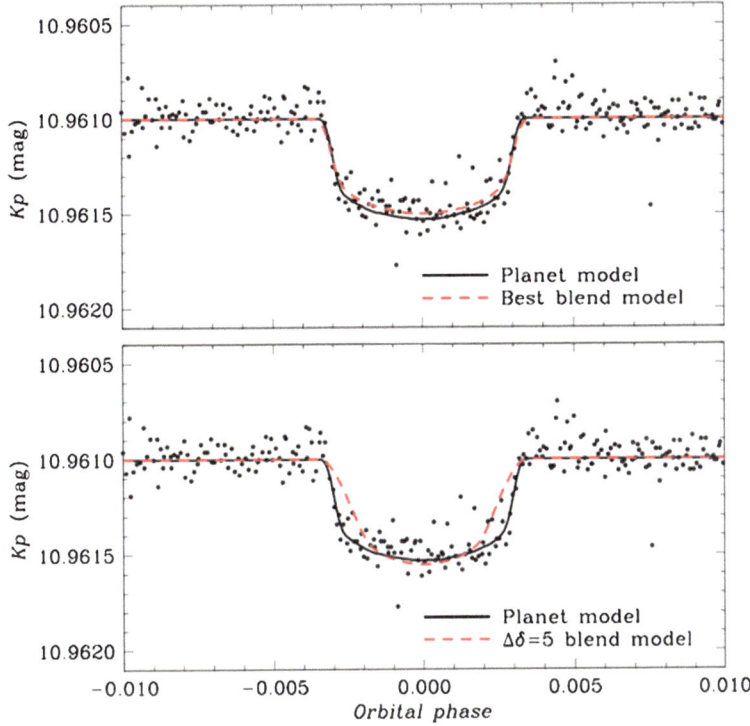

Fig. 24.4 Light curve for planet Kepler-10c (Credit: NASA)

Secondly, it is possible for amateurs to gather their own light curves. This requires a telescope fitted with an accurate guiding system that can fix onto one star and stay there, with a CCD camera to gather the data, which can then be analyzed via various pieces of software. The results are impressive, comparing well to the light curves established via professional instruments. Using a 6-cm refractor and an 8-bit camera, it was possible for Michael Theusner to create this light curve for the star HD189733. Who knows where such work might lead?

The Kepler website at http://kepler.nasa.gov/ provides a portal to all the spacecraft information, while http://www.planethunters.org offers the chance to make a real contribution via the web. If you want to create your own light curves, try http://www.transitsearch.org/ and look at http://astronomyonline.org/Exoplanets/AmateurDetection.asp.

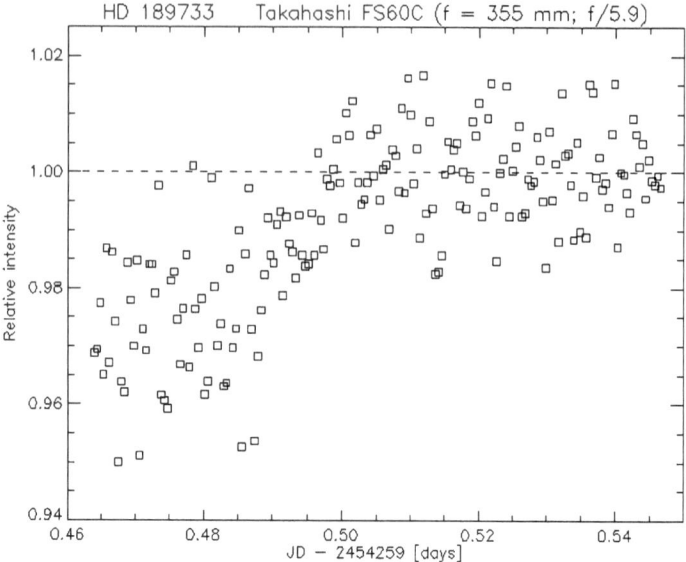

Fig. 24.5 Results gained by Michael Theusner for the star HD189733 (Credit: Michael Theusner, www.theusner.eu with www.theusner.eu/downloads/esop2008_theusner.pdf for details on data collection)

Is This the One?

A sense of future possibilities and the further shrinkage of the size of exoplanets that can be identified was provided in late 2011 by an announcement from the Kepler mission science team. They offered evidence for the existence of two Earth-sized planets around a star identified as Kepler-20. Already well supplied with an outer solar system of three large planets, Kepler 20e and 20f are, respectively, 1.03 times larger and 0.87 times smaller than our home.

The evidence for these discoveries comes via the transit method, the now familiar decrease in observed starlight in proportion to a planet's radius. The result is not yet certain. This is the outer rim of the exploration of deep space, and the possibility of false positives remains. Yet there is enough evidence to suggest, if nothing more, that the planet Kepler 20f could have a composition similar to that of Earth and an atmosphere thick with vapor.

Are we looking at a solar system analogous to our own? Is Kepler-20f a water-laden world? The Kepler mission has transformed our understanding of stars and their systems. It raises new questions all the time. What are the possibilities for life? Is liquid water common? Does the Drake equation still stand? We want to discover whether or not we are alone, and we will not rest until we know.

Chapter 25

Watching the Sunrise

Observation: Altai Scarp, Craters Clavius and Bailly
Significance: Effects of the Moon on Earth, what we do not know
Science: Far side of the Moon, South Pole, search for water

In this last chapter, we turn our telescopes back to the Moon. Small when compared to some of the giant planets' satellites, the Moon is our intimate companion and has played a vital role in the history of our planet. It probably restrained the young Earth from developing too great an orbital spin. It determines the cycle of the tides. If computer simulations are correct when they indicate that one in 12 terrestrial planets may host a moon something like ours, then these moons may be signposts pointing to life in the depths of space.

Our Moon is not unique, but it is special. Our observing takes us to the edges of the Moon, to the shrunken perspectives of the lunar south and the rim of the Aitken Basin, as well as to will o' the wisp features, sometimes there and sometimes not, granted to us by libration. We return to Galileo, who having observed the Moon, chose to ignore it as he marshaled his arguments for the Copernican universe, and also to Newton, for whom the Moon remained a puzzle.

The Wild South

We return to the Moon where we left it, among the scarps and straight lines that look from Earth like sheer cliffs. They are striking observing targets. One of the best is seen on a 5-day old Moon, starting from the center of the crescent and moving south. Binoculars will reveal first the familiar outline of the crater Theophilus, just emerging into a shadowland dawn from lunar darkness, and then the Rupes

M. Marett-Crosby, *Twenty-Five Astronomical Observations That Changed the World: And How To Make Them Yourself*, The Patrick Moore Practical Astronomy Series, DOI 10.1007/978-1-4614-6800-4_25, © Springer Science+Business Media New York 2013

Altai or Altai Scarp, a wall stretching from northeast to southwest along the surface of the Moon. The wall is impressive but short-lived, its shadows swallowed by lunar day. If you miss it first time around, then it reappears some 4 days after full.

The Altai is a huge feature, some 480 km long and as much as 3 km high. It is probably the last evidence of the massive impact that created the Mare Nectaris, which we observed in a previous chapter. The Altai is its outermost crater wall. The great age of this impact is revealed by a quick crater count to left and right and even on the scarp itself.

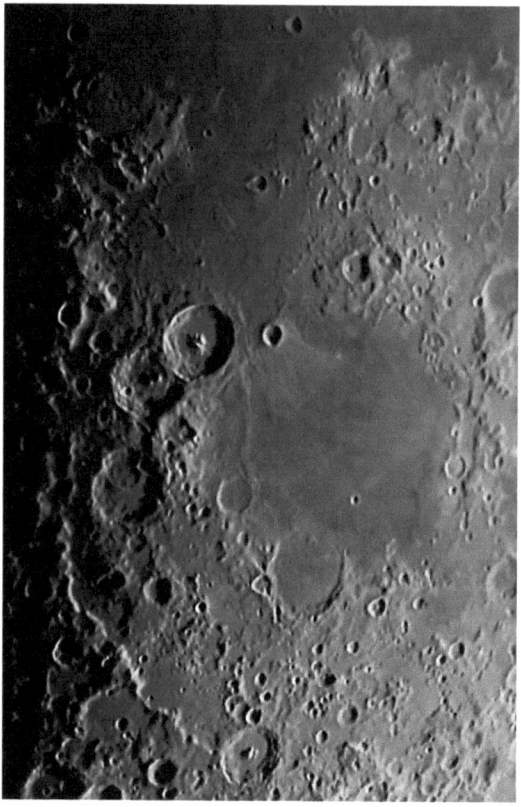

Fig. 25.1 Mare Nectaris imaged on 1 November 2011; a mosaic of 7 images from stacked video taken afocal on Canon Digital Ixus 800. Stacked in Registax. 10 mm eyepiece; 2x Barlow; Skywatcher 150 Explorer Newtonian. Stitched in photoshop. Wavelets in Avistack 2 (Credit: Julian Cooper)

Follow the Altai to its southwestern end and you will reach Piccolomini. One of several craters in the lunar south named after supporters of the Ptolemaic universe, Piccolomini lies some 200 km south of a crater we observed earlier, Fracastorio. Indeed, it is worth drifting north from Piccolomini to refresh the memory as to the range of features that are covered by the catch-all label 'crater.' Fracastorio is a flooded ghost while Piccolomini is some 4 km deep with terraced crater walls, hills and a complex central massif. Through a telescope, Piccolomini makes for one of the finest of all lunar crater observations.

Alessandro Piccolomini (1508–1579) was an Italian bishop, classicist and astronomer whose 1540 work *The Fixed Stars* contained a catalog organized by constellation and an early classification of the stars, each with a label in order of brightness. His was the first such scheme, anticipating by 63 years that of Johann Bayer. Piccolomini's other work, *On the Sphere,* presents a vigorous defense of the Ptolemaic system.

Beyond Piccolomini and towards the pole lies a wild, chaotic region, the Southern Highlands. It is austere and complex, a witness to the violence of the lunar past. By day eight of the Moon, the principal feature of this area is emerging from its 2-week night, the crater Clavius. Since this is a difficult area to navigate with certainty, the best place to start is the Tycho Crater. Clavius is the large crater south of Tycho, above it if you are observing through a telescope. A naked-eye notch on the line of the terminator, it makes for a fantastic telescopic target and a strange one, because its shape is never quite revealed. One of the largest craters on the Moon, the effect of curvature makes it appear stretched rather than round. Nevertheless, it is easy to peer inside and follow the alignment of the craters that arc across its floor. Particularly striking are the two craters embedded in Clavius' decayed walls – both Porter and Rutherford have central peaks of their own, while those within the bowl are neat punches into the surface, a reminder of the difference between simple and complex craters.

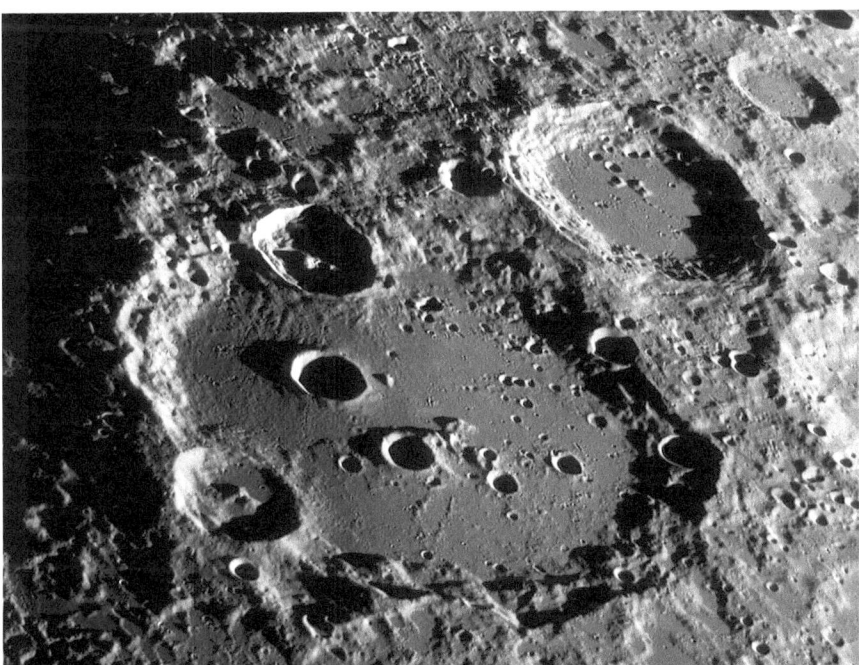

Fig. 25.2 Clavius, imaged on 26 July 2008 using a Celestron 14 in. Schmidt-Cassegrain telescope, a red filter and a Skynyx 2.1M video camera. The exposure time was 25 milliseconds and the image is the result of the addition of 800 video frames (Credit: Christian Viladrich (http://christian.viladrich.perso.neuf.fr)

The End of the Old World

We can pause here to recall the crater's patron, Christopher Clavius (1538–1612), another defender of the Ptolemaic system remembered on the Moon. Clavius was among the last astronomers to write a commentary on John of Sacrobosco. In this sense, Clavius belongs to the medieval world we explored in Chap. 1. We know why he came to be interested in astronomy because he writes about it himself, describing how, as a 22-year-old studying in Portugal, he was outside at midday and able to observe: "The Moon placed between my sight and the Sun with the result that it covered the whole Sun for a considerable length of time. There was darkness in some manner greater than night; neither could one see where one stepped. Stars appeared in the sky and (marvelous to behold) the birds fell down from the sky to the ground in terror of such horrid darkness."

This eclipse inspired him for his whole life.

Clavius met Galileo, and as a old man he lived to see the publication of *Sidereus Nuncius*. He notes how a long tube-shaped instrument arrived in Italy that "shows many more stars in the firmament than can be seen in any way without it." When turned towards the Moon, the effect of this new telescope is even more remarkable: "When the Moon is a crescent or half full, it appears so remarkably fractured and rough that I cannot marvel enough that there is such unevenness in the lunar body. Consult the reliable little book by Galileo."

We can hear the world changing in these words. It is as if Clavius, an immensely respected Jesuit mathematician, knows that the Aristotelian universe that has shaped his view of everything is crumbling. He has the grace, indeed, to point to the new astronomy as a more reliable guide.

Luna Incognita

As the angle of sight foreshortens around Clavius, so our telescopes approach the south pole of the Moon wherein lies the Aitken Basin. Altogether invisible from Earth, it was *luna incognita*, an unknown land, until imaged by the Clementine probe in 1994.

The mission got its name from the song "My Darlin' Clementine." Clementine in the ballad is the daughter of a miner, and the mission was due to explore the mineral content of the Moon. Also, once its mission was done, the spacecraft would be "lost and gone forever," as the song says.

Clementine was the first U. S. lunar mission for more than two decades after Apollo, and during its 71 days in orbit it mapped 38 million square km of the Moon in 11 visible and near-infrared colors, some 940,000 images in total. It gave us a first sight of the lunar south pole.

Fig. 25.3 The south pole of the Moon as imaged by Clementine (Credit: NASA/JPL/USGS)

Aitken is the oldest known impact basin in the Solar System, and also one of the largest. First theorized to exist in the 1960s by William Hartmann and Gerard Kuiper – they called it the 'Big Backside Basin' – Clementine data reveals a vast crater nearly 8 miles deep. It is a window into the inner structure of the Moon, with rocks that have a different composition to those found elsewhere, revealing the nature of the deep crust and perhaps even the mantle. Surrounded by peaks and crater rims, the south pole region includes many other craters, for example Schrödinger to the lower right of the above image. It is here, in these lands of permanent shadow, that Clementine also detected the signature of water. Crater walls were more reflective than they should have been, a sign of ice within the regolith.

Recent missions to the Moon have expanded our knowledge on the presence of lunar water. The M3 instrument on board *Chandrayan 1* identified the signs of water spread widely but thinly across the lunar surface, and a re-analysis of sediments returned to Earth by *Apollo 17* uncovered beads of volcanic glass containing a hundred times more water than was measured the first time around. The Cassini probe also detected water on its way to the Saturn system. When LCROSS slammed its projectile Deep Impact into the lunar surface on October 9, 2009, it was established from the ejecta that water makes up some 0.5 % of the whole Moon.

Where does this lunar water come from? As on Earth, the Moon might have gained water from asteroids and comets – the latter being more likely. Some water might also be a by-product of the solar wind.

NASA's Lunar Crater Observing and Sensing Satellite (LCROSS) was launched in June 2009 with the specific goal of confirming the presence or absence of water ice at the lunar south pole. It fired an impactor into the Cabeus crater on October 9. The orbiting satellite then flew through the debris taking measurements. Careful observations of the debris were also made from Earth.

What are the implications of this wetter Moon? Claims that a lunar base could mine this water and thus overcome one of the hurdles to human occupation may be exaggerated, especially if, as seems possible, the water is laced with heavy atoms such as mercury and silver. But deep places such as the Aitken Basin are a locked library, storing evidence of the earliest history of Earth. Sealed in there are records of very early collisions and of lunar volcanism. There might even be evidence of the early Earth's atmosphere to be found. A sample-return mission to the south pole of the Moon could revolutionize the history of the Solar System and Earth and answer some of the many questions left open in this book.

Newton's Puzzling Moon

The Moon is so close to us that we think we must know it well. "It is a province of the Earth, a mere stone's throw away," declares the Flammarion Book of Astronomy in its 1964 edition, adding that its "insignificant astronomical distance is hardly worthy to be called astronomical." Yet Isaac Newton in his great *Principia* could not explain its motions fully.

This was not something Newton himself admitted, but his translator John Machin (c. 1686–1751) acknowledged that Newton's account was insufficient. "It seems that there is more force necessary to account for the motion of the Moon's apogee."

Where was that "more force" to come from? Was the Moon to prove Newton wrong? Alexis Clairaut (1713–1765) proposed to the Royal Academy of Paris that Newton's laws of gravity would have to be amended to take into account the nearest astronomical object to our shores, but while this was being debated he made new calculations that, when incorporating the gravitational effects of Earth as well as the Sun upon the Moon, enabled him to not to reject but to provide new evidence of the rightness of Newton's theory. His *Theorie de la Lune* of 1750 was a clever approximate solution to that most vexing of mathematical conundrums, the three-body problem, and he tied the Moon into Newton's system. Leonhard Euler (1707–1783) wrote in 1752 that, "Mr. Clairaut has made the important discovery that the movement of the apogee of the Moon is perfectly in accord with the Newtonian hypothesis." All was well with the heavens after all.

Beyond Clavius, the southern Moon is untamable through a telescope. The angles are all wrong; everything is shrunken. As you move further towards the pole, the Moretus crater stands out, but it is a tricky observation. Beyond Moretus lies the Bailly crater.

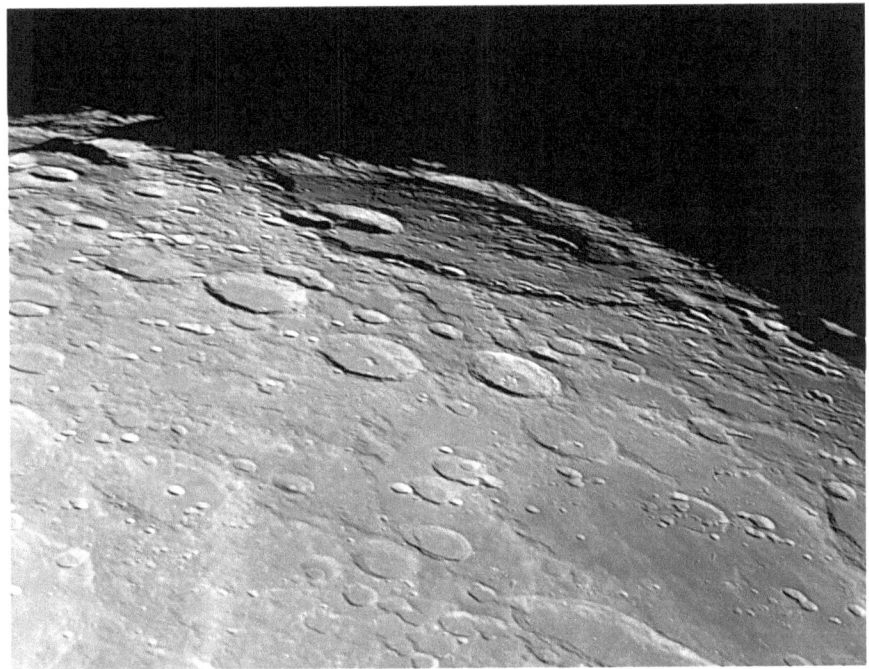

Fig. 25.4 Bailly crater, imaged on December 22, 2007 with a Celestron C11 telescope (11 in. Schmidt-Cassegrain) was used, on a SkyWatcher HEQ5 equatorial mount; Camera a DMK 31AF03 (Credit: Peter van de Haar, http://www.footootjes.nl)

To find Bailly, wait until the Moon is about 90 % towards full. Use the Tycho crater and its rays as guideposts, following the ray patterns towards the south pole. A pretty quartet of perfect lunar craters lie before a high crater rim. Looking beyond that reveals not so much a crater as a rampart-bound plain, vast and still, always sliding over the edge of vision into darkness.

The Bailly crater lies on the rim of the southwestern lunar limb as it faces us. From Earth, therefore, Bailly is always seen at an oblique angle that foreshortens the appearance of what is in fact one of the largest craters on the lunar surface, 183 miles across and with crater walls 14,000 ft high in places. It is a field of ruins punctured by smaller craters, the largest of which (Bailly A and Bailly B) are visible under the right conditions. Bailly is one of the oldest visible lunar formations.

How is it possible even to see Bailly? Our view of the Moon is not constant. As much as 59 % of the lunar surface is visible from Earth, the result of an effect called libration. Planetary scientists distinguish three kinds of libration by which the Moon at times shows more than half its face – eccentricities in the Moon's orbit of Earth, variations in the inclination between the Moon's axis of rotation to the plane of its orbit and the effects of Earth's rotation and our place on the planet's surface. We can distinguish more simply between optical and physical librations – those to do with the Moon and those to do with where we are looking from. The effect is that the Moon's south pole tips towards us. We can glimpse around the corner to the far side of the Moon.

The word libration is derived from the Latin *libra*, meaning scales, and libration refers to a motion comparable to the swaying backwards and forwards of weighing scales. John Flamsteed applied it to the Moon around 1670.

It is perhaps suitable that this swinging crater, sometimes there and sometimes not, should have acquired the name of an astronomer whose fortune swung from fame to infamy. Jean Sylvain Bailly (1736–1793) was a leading light of eighteenth-century Paris who wrote knowledgably on the satellites of Jupiter and creatively about much else. In the early years of the French Revolution, he was a parliamentarian and mayor of Paris, political achievements for which he was executed on November 12, 1793. He is one of very few victims of the guillotine to find a sort of immortality upon the Moon.

Fig. 25.5 The Far Side of the Moon, imaged by Apollo 16 in April 1972 (Credit: NASA)

The 41 % of the Moon permanently turned away from Earth was first photographed by the Soviet *Luna 3* probe on October 7, 1959. Its 18 original images were augmented a year later by a further 25 from the camera on board another Soviet craft, *Zond-3*. Frank Borman, James Lovell and William Anders were the humans to look directly upon its surface in December 1968. Since then, other Apollo missions – the above image was captured by *Apollo 16* in April 1972 – and unmanned probes have passed over its hidden side.

The far side of the Moon is very different from the nearside. It lacks maria, presenting a scene much like the near side highland terrain we observed towards the south pole. It is very heavily bombarded, and the Clementine mission established that lunar crust on the far side is substantially thicker than on the near. NASA's Lunar Prospector Mission identified unusual surface heavy elements including thorium, titanium and iron.

Why this asymmetry? The effects of impacts on the near side and of Earth's gravity have been suggested as causes. It has also been suggested that ejecta from a very early near side basin had an effect on the topography and crustal thickness of the far side.

It is also possible that the far side of the Moon is evidence for how the Moon was made. As we discussed in an earlier chapter, one theory with a lot of support suggests that the Moon is the result of a catastrophic impact upon the young Earth. Scientists have postulated that this collision led to the formation of both the Moon and another body. This second object collapsed onto the far side, coating it with the extra crust. This may be all that remains of Earth's second Moon.

There is much we do not fully understand about our nearest neighbor in space. Some of its secrets will be revealed by the NASA Gravity Recovery and Interior Laboratory (GRAIL) mission, a twin spacecraft survey that will produce a detailed and very precise map of the Moon's gravitational field. The probes are able to measure changes in the distance between them down to the width of a single red blood cell. The resulting data may well provide insights into the Moon's thermal history and interior. Meanwhile two ARTEMIS satellites, repurposed when their earlier mission ended, are now in orbit around the Moon exploring the solar wind in the Moon's 'wake' as well as crustal magnetism and the lunar interior.

The Moon's Dominion

The Moon leaves its mark on the planet it orbits. One of these might be life. Over the history of the Earth-Moon system, the satellite has had a stabilizing influence upon the axial tilt and rotation of Earth and therefore upon the climate. Some theorists of life on exoplanets even propose that the presence or absence of a moon should be included within the Drake equation. But the most apparent daily footprint of the Moon upon Earth is the tide.

If we imagine first that Earth was entirely a water-world and second that the Moon sat still above the equator, then it is fairly clear that, as this wet Earth rotated on its axis, the water would be pulled into the shape of an oval, with a bulge reaching out towards the Moon powered by gravitational attraction. This is called the direct tide. Of course, the fact that Earth is not completely covered in water makes the situation more complex. So does the Moon's own rotation. There are in fact two bulges, one facing the Moon and another on the opposite side of Earth, where the gravitational attraction is weakest. Earth is therefore subject to two tidal movements in a cycle that lasts about 24 h and 50 min, which is the length of time the Moon takes to return to the same position with respect to an observer on Earth.

The phenomenon of the gravity of Sun and the Moon pulling at the oceans in unison is called syzygy, derived from Greek for "yoked together," as in cattle pulling a plough. The opposite of syzygy is quadrature, when the Moon and the Sun are acting at right angles and their influences work against each other.

The movement of Earth's water is also affected by the gravity of the Sun. When Sun and Moon are aligned with Earth, the Sun's gravity works with that of the Moon to produce higher tides or springs.

Although we now take for granted that the Moon causes the tides, the evidence is not obvious. Indeed, Galileo himself dismissed the lunar bond with the tides when in 1632 he produced his *Dialogue Concerning the Two Chief World Systems*, his most thorough-going defense of the Copernican universe. "Spare us the rest," he writes when the speaker Simplicio proposes to talk about the Moon's impact on the sea. "I do not think there is any profit in spending the words to refute them."

Why does Galileo dismiss it so quickly? As we have seen, there were alternative theories that explained the observations made by early telescopes. They could be shoe-horned into the Ptolemaic system or fitted into either the Copernican or Tycho Brahe's model. Galileo knew that the phases of Venus and the movement of the moons around Jupiter could both be explained by Tycho's system. So, even, could sunspots. In the *Dialogue*, Galileo points out the deficiencies of the Ptolemaic account and refers to the sunspots, but he needed a physical proof of what he believed was true, that Earth was in motion around the Sun.

He chose the tides, arguing that the ebb and flow of the sea proved the motion of Earth. It was the third of his "very convincing evidences" and was meant to be the knockout blow.

It did not convince even then. The papal commission that read a draft of the book singled out this point as failing to fit the evidence. A correspondent wrote to Galileo in 1632 to say, "I marvel…at the fact that you leave an important question unanswered, namely that the tide should occur every day at the same time, whereas common opinion holds the contrary." High tide, after all, takes place twice a day and not at the same time.

These were pretty good counter-arguments. Staunch defenders of Galileo such as Pierre Gassendi (1592–1655) found them impossible to parry. It did not help that Galileo, characteristically forthright, had described the lunar theory as ridiculous, noting that even Kepler, "Despite his open and acute mind…has nevertheless lent his ear and his assent to the moon's dominion over the waters." Uncomfortable with Tycho's system and dismissive of the Ptolemaic, he needed direct observational evidence that Earth went around the Sun and thought he'd found it. As J. L. Heilbron expresses it in his biography *Galileo* (Oxford, 2010), "This was the biggest bluff of Galileo's career."

He had looked elsewhere for the missing evidence. It has been demonstrated in recent years that Galileo split the double star Mizar in 1617 and then attempted to use the observations to demonstrate parallax. He did the same a little later among the stars of the Trapezium. When they did not help him, he fell back upon the tides.

At the end of the night, as dawn touches the eastern horizon, the last quarter Moon is low in the sky. It's worth making an early start and, armed with telescope and coffee, observe this unfamiliar phase. Two great craters slide across into the darkness just north of half way up the crescent. The one further into the light is Copernicus, and it is preceded by Eratosthenes. This last is a must-see lunar crater either now or on the face of an 8-day Moon, 58 km broad and positioned at the western end of the Apennines. It is deeper than its neighbor Copernicus and surrounded by fine ridges. It stands on the border between two stretches of lunar maria.

This image of Eratosthenes was taken from the Command and Service Module of *Apollo 17* in December 1972, Ronald E. Evans piloting the spacecraft. The Copernicus crater is sliding over the rim, while Eratosthenes and its surrounds dominate the landscape.

Fig. 25.6 Eratosthenes crater, taken looking southward from the command and service module (Credit: NASA)

The last days of the old Moon are dominated by sunset over the Oceanus Procellarum, the frozen lava ocean of silent storms. To the south of this great stretch lies the Gassendi crater and then the Mare Humorum. This lunar sea betrays more obviously than others its origin as an impact crater. Gassendi is straightforward to spot on the rim, and in a telescope with good magnification it's possible to see the way in which the interior of the crater is higher than the surrounding terrain and the complex pattern of fault lines. Look out, too, for the later impact that has shattered its north wall.

And then nothing. Night swallows the Moon and our own day engulfs it. Always, though, to begin again.

There is much still to discover about the Moon. Note that even where we have touched another world with our own hands, we still do not know everything. If there is one simple message that emerges out of astronomy, it is that there is always more to see and more to understand.

And we, as amateur astronomers, can share in the learning process. To repeat these 25 groups of observations is to visit some remarkable sights in the sky and to explore some fascinating aspects of modern science. It is also to become connected to moments in history when whole patterns of thoughts have been overturned. New astronomical discoveries have made Earth less central and the heavens ever more vast.

And the next, 26th observation to change the world? It might be one of many things. Perhaps the identification of exoplanets will demand of us that we start again our effort to explain the Solar System and the formation of rocky planets like Earth. The Mars Science Laboratory's Curiosity Rover might detect water on the surface of that planet and even the signatures of life. We might find a way to observe and interact with dark matter or glimpse the hidden secrets of black holes. The New Horizons mission to Pluto and the Kuiper Belt will surely reveal much that we do not know about the outer reaches of the Solar System. Some time, quite soon now, the Voyager mission will break out of the Sun's grasp and enter interstellar space. The search for extraterrestrial intelligence might show that we are not alone. An amateur astronomer could spot an inbound asteroid.

Whichever, the journeys we make to the stars will continue to change us.

About the Author

Michael Marett-Crosby obtained an M.A. in History and a D.Phil from Oxford University. His published books include *The Conversion of England* (Abbey Press 1998) and *Doing Business with Benedict* (Continuum 2002). He speaks and writes on astronomical subjects, and believes practical astronomy is important in building bridges between science and the liberal arts.

A lifelong student of astronomy, he has been an enthusiastic amateur astronomer for most of his life, and is fortunate to be able to view the night sky from the enviable viewpoint of the UK's southernmost outpost in Jersey. He is lead trustee of a charitable trust.

This book is dedicated to Ollie, who looked into the sky one evening and saw space.

M. Marett-Crosby, *Twenty-Five Astronomical Observations That Changed the World: And How To Make Them Yourself*, The Patrick Moore Practical Astronomy Series, DOI 10.1007/978-1-4614-6800-4, © Springer Science+Business Media New York 2013

Index

A

Abell, George Ogden (1927–1983), 257
Adams, Douglas (1952–2001), 263
Adams, John Couch
    (1819–1892), 224
ALH84001, 87
Amalthea, moon of Jupiter, 44
Apollo NASA missions
    Apollo 8, 123
    Apollo 11, 5, 121, 122
    Apollo 16, 307
    Apollo 17, 10, 122, 304, 309
Arecibo Blast, 137
Argelander, Freidrich Wilhelm
    (1799–1875), 153
ARTEMIS, NASA mission, 307
Asimov, Isaac (1920–1992), 295
2001 A Space Odyssey, 5, 235
Asteroids
    (1) Ceres, 192, 194–196
    (433) Eros, 199
    (951) Gaspra, 198
    near-Earth, 197, 199, 201, 203
    observing, 193, 195, 196, 200
    (4,179) Toutatis, 194
    types, 199
    (4) Vesta, 194, 196, 199
Aurora, 23, 26, 83, 172, 221

B

Baade, Walter (1893–1960), 185
Banks, Joseph (1743–1820), 269
Barnard, E.E. (1857–1923), 94
Barringer Crater, Arizona, 118, 201
Bayer, Johannes (1572–1625), 62
Bayeux Tapestry, 151
Benjamin, Walter (1892–1940), 101
Bessel, Freidrich (1784–1846), 110
Black holes
    Cygnus X-1, 231
    formation, 232
    and galaxies, 234–235
    Sgr A*, 234, 235
Boscovich, Roger (1711–1787), 125
Brahe, Tycho, 12, 77, 110, 174, 176, 192, 205,
    207, 215, 308
Brown dwarf stars, 176, 181
Burnell, Jocelyn Bell (born 1943), 212, 213

C

Callisto, moon of jupiter, 41, 47
Cannon, Annie Jump (1863–1941), 30
Cassini, Giovanni Domenico
    (1625–1712), 242
Cassini, NASA mission, 242–245, 247, 248
Castelli, Bendetto (1578–1643), 32

M. Marett-Crosby, *Twenty-Five Astronomical Observations That Changed the World:*   313
And How To Make Them Yourself, The Patrick Moore Practical Astronomy Series,
DOI 10.1007/978-1-4614-6800-4, © Springer Science+Business Media New York 2013

Cepheid variables, 72, 158, 160, 162, 185, 253
Chandrasekhar limit, 109, 208
Chandrasekhar, Subramanyan (1910–1995), 109
Chandrayan lunar probe (ISRO), 117
Clairaut, Alexis (1713–1765), 149, 304
Clarke, Arthur C. (1917–2008), 295
Clavius, Christopher (1538–1612), 302
Clementine, NASA mission, 303, 307
Clusters
    globular, 61, 66, 179, 183–186, 263, 291
    open, 62, 106, 179, 181, 183, 185,
        187–189
Coleridge, Samuel Taylor (1772–1834), 226
Comets
    observing, 139–142
    structure, 142
    sungrazers, 143, 144
    types, 143, 146
Constellations
    Canes Venatici, 233, 234
    Cassiopeia, 70, 186, 205
    Cepheus, 158, 229
    Cetus, 162, 164, 254
    Corona Borealis, 183, 207
    Cygnus, 64, 211, 229, 234
    Eridanus, 129–131, 134, 136, 138
    Gemini, 287, 288
    Orion, 42, 53, 54, 56, 104
    Perseus, 154, 188
    Sagittarius, 66, 67
    Ursa Major, 27–29, 34
    Virgo, 65, 185, 257, 258
Copernicus, Nicholas, 76
Cosmic Microwave Background (CMB),
        136, 261
Curtis, Heber Doust (1872–1942), 72

D

Dark Matter, 74, 253, 260, 261, 310
D'Arrest, Heinrich Louis (1822–1875), 224
da Vinci, Leonardo, 7
Deep Sky Objects (not Messier)
    Double Cluster (Caldwell 14), 187
    Horsehead Nebula (IC434), 94
    Hyades (Melotte 25), 181
    Melotte 20, 155, 156, 188
    NGC1232, 134
    NGC1300, 134, 135
    NGC1528, 189
    NGC1535, 133, 134
    NGC2392, 289
    NGC3718, 256

    NGC3729, 256
    NGC4151, 256
    NGC6229, 184
    NGC7000, 230
    NGC7662, 70
    NGC7789, 187
    Veil Nebula (NGC 6960,6992,6995), 211
Dickens, Charles (1812–1870), 177
Donnolo, Shabbetai (born 913), 37
Don Quixote, ESA mission, 203
Drake Equation, 129, 137, 294, 298, 307
Drake, Frank (born 1930), 130
Draper, Henry (1837–1882), 98
Dryer, John Louis Emil (1852–1926), 50

E

Einstein, Albert (1879–1955), 275
Enceladus, moon of Saturn, 239
Escape velocity, 232
Europa, moon of Jupiter, 41
Exoplanets
    direct observation, 308
    Doppler spectroscopy detection, 289–290
    radial velocity detection, 287
    transit detection, 293

F

Fraunhofer, Joseph (1787–1826), 17

G

Galaxies
    Seyfert, 253, 256
    structure, 69
Galilean moons. *See* Io, Europa, Ganymede,
        Callisto
Galileo Galilei (1564–1642), 6
    and the moon, 42, 43
    and moons of Jupiter, 15, 39, 43
    and sunspots, 20
    and tides, 308
Galileo, NASA mission, 173
Galle, Johann Gottfried (1812–1910), 224
Ganymede, moon of Jupiter, 2, 248
Genesis, NASA mission, 24
Giotto di Bondone (c1267-1337), 151
Giotto, ESA mission, 146
Goodricke, John (1764–1786), 156
GRAIL, NASA mission, 307
Gregory, James (1638–1675), 110, 269
Gruithuisen, Franz (1774–1852), 276

H

Halley, Edmund (1656–1742), 148
Halley's Comet, 146
Hardy, G.H. (1877–1947), 250
Hardy, Thomas (1840–1928), 236
Harriot, Thomas (c1560-1621), 42
Hayashi Tracks, 96
Helium, 17–18, 33, 61, 66, 131, 159, 170–172,
        208, 209, 221, 222, 233, 234, 275, 290
Herschel, Caroline (1750–1848), 172
Herschel, John (1792–1871), 135
Herschel, William (1738–1822), 17
Hertzsprung-Russell diagram, 53–55, 96,
        108, 186
Hevelius, Johannes (1611–1687), 148
Hewish, Antony (born 1924), 212
Hornsby, Thomas (1733–1810), 269
Hubble Constant, 175, 257
Hubble, Edwin (1889–1953), 72
Hubble's Fork, 73
Hubble Space telescope, 57, 65, 99, 100, 108,
        119, 161, 164, 165, 169, 185, 195, 197,
        221, 222, 246-247, 254, 257, 260,
        291, 292
Huygens, Christian (1629–1695), 110, 242, 245
Huygens, ESA lander, 243

I

Individual Stars
        Achernar (Alpha Eridani), 136
        Adhara (Epsilon Canis Majoris), 105
        Alderamin (Alpha Cephei), 160
        Algol (Beta Persei), 188
        Alioth (Epsilon Ursae Majoris), 32
        Alkaid (Eta Ursae Majoris), 29
        Alnilam (Epsilon Orionis), 57
        Alnitak (Zeta Orionis), 57, 59
        Alpheratz (Alpha Andromedae), 68
        Arcturus (Alpha Boötes), 132
        Barnard's Star, 33
        Beid (Omicron 1 Eridani), 130–131
        Bellatrix (Gamma Orionis), 56, 59
        Beta Pictoris, 292, 293
        Betelgeuse (Alpha Orionis), 56
        Blaze Star (T Corona Borealis), 207
        Caph (Beta Cassiopeiae), 186
        Castor (Alpha Geminorum), 289
        Cor Caroli (Alpha Canum Venaticorum), 233
        Cursa (Beta Eridani), 129
        61 Cygni, 110, 111, 132
        Delta Cephei, 158–160
        Deneb (Alpha Cygni), 111
        Dubhe (Aha Ursae Majoris), 35
        Epsilon Eridani, 131–133
        Fomalhaut (Alpha Piscis Australis), 288
        Furud (Zeta Canis Majoris), 107
        Garnet Star (Mu Cephei), 159
        Hind's Crimson Star (R Leporis), 60
        Kaus Borealis (Lambda Sagittarii), 66
        Keid (Omicron 2 Eridani, 40 Eridani), 131
        La Superba (Y Canum Venaticorum), 233
        Megrez (Delta Ursae Majoris), 35
        Mintaka (Delta Orionis), 57
        Mira (Omicron Ceti), 162, 164
        Mirzam (Beta Canis Majoris), 104
        Mizar (Zeta Ursae Majoris) and Alcor, 31
        51 Pegasi, 287, 290, 291
        Pollux (Beta Geminorum), 289
        Porrima (Gamma Virginis), 258
        Regulus (Alpha Leonis), 258
        Rigel (Beta Orionis), 59
        Sirius (Alpha Canis Majoris), 103, 104
        SN1987A, 205, 208, 209
        Trapezium stars (Orion Nebula), 97
        UV Ceti, 131, 133
        Vega (Alpha Lyrae), 183, 229
        Vindemiatrix (Eta Virginis), 259
        VY Canis Major, 105, 106
International Space Station
        history, 277–279
        observing, 279–280
Io, moon of Jupiter, 41
Iridium flares, 277, 280

J

Jansen, Pierre Jules (1824–1907), 18
Jeans, James (1877–1946), 95
Johnson, Samuel (1709–1784), 236
Jupiter
        Great Red Spot, 169, 172, 173
        observing, 167–170
        and the speed of light, 167, 175, 176
        structure of gas giants, 167

K

Kepler, Johannes, 6, 12, 126, 173, 214
        Somnium, 126, 127
Kepler, NASA mission, 8, 296–298
Kirchhoff, Gustav (1824–1887), 17
Kuiper Belt, 2, 122, 133, 145, 191, 198, 310

L

Lacaille, Nicholas de (1713–1762), 138
Lassell, William (1799–1880), 218

LCROSS, NASA mission, 304
Leavitt, Henrietta Swann (1868–1921), 72, 160
Le Verrier, Urbain (1811–1877), 224
Ley, Willy (1906–1969), 277
Lockyer, Norman (1836–1920), 18
Lomonosov, Mikhail (1711–1765), 268
Lovecraft, Howard Philips (1890–1937), 165
LRO. *See* Lunar Reconnaissance Orbiter
        (LRO)
Lunar Reconnaissance Orbiter
        (LRO), 5, 121, 122

M
Maestlin, Michael, 12
Magellan, NASA mission, 270, 271
Magnitude, apparent and absolute, 31, 153,
        160, 230, 241, 250
Mariner 2, NASA mission, 23
Mars
    life on, 84, 85, 87–88, 295
    moon Phobos and Deimos, 78
    observation, 75, 77
    orbit, 75
    rovers spirit and opportunity, 86
    science, 87, 310
Maskelyne, Nevil (1732–1811), 218, 227
Maurolyco, Francisco (1495–1575), 97
Maury, Antonia (1866–1952), 30
Maxwell, James Clerk (1831–1879), 245
Mayr, Simon (1573–1624), 43
Mercury, 1–30, 33, 38, 41, 42, 60, 68, 79, 101,
        162, 191, 192, 197, 248, 265, 266,
        268–270, 272–276, 304
Messenger, NASA mission, 2, 274
Messier, Charles (1730–1817), 36
Messier objects
    M1 (Crab), 210
    M8, 66
    M13, 183, 184
    M24, 67
    M31 (Andromeda Galaxy), 68, 253
    M32, 69, 70
    M34, 156, 188
    M41, 106
    M42, 93
    M43, 93, 94
    M44 (Beehive), 78, 181–183
    M45 (Pleiades), 95
    M51 (Whirlpool), 233, 234
    M60, 259, 260
    M77, 254–256
    M79, 61

M81, 36
M82, 36
M87, 185, 260
M88, 262
M91, 262
M92, 184
M104, 65
M106, 233
Meteor showers, 142, 143
Micromégas, 113, 114
Milky Way, 42, 63–69, 71, 72, 74, 161, 166,
        185, 186, 208, 213, 230, 253, 262, 263,
        291, 294
Minor Planet Center, 194, 195, 201
Mir, Soviet / Russian space station, 282
Mitchell, John (1724–1793), 231
Moon, in general
    craters, simple and complex, 301
    and Earth, 1–6
    Earthshine, 7–8
    far side, 117, 299, 305–307
    librations, 306
    orbit, 6, 306
    origins, 125
    phases, 1, 6, 7
    syzygy and quadrature, 308
    and tides, 307–309
    water, 1, 7, 8
Moon Mineralogy Mapper (M3), 117, 303
Moon, named features
    Altai Scarp, 299, 300
    Bailly crater, 305
    Catharina crater, 117
    Clavius crater, 5, 299, 301
    Copernicus crater, 1, 10, 309
    Cyrillus crater, 117
    Eratosthenes crater, 124, 309
    Fracastorius crater, 120
    Kepler crater, 9
    Mare Nectaris, 5, 119–121, 300
    Piccolomini crater, 300
    Pictet crater, 5
    Rupes Recta, 124, 125
    South Pole Aitken Basin, 99, 302, 304
    Theophilus crater, 116–118
    Tycho crater, 4, 10, 301, 305

N
Nebra Sky Disk, 189
Nebula, types of, 95, 99
Neptune
    discovery, 178, 192, 217

observing, 226
Triton, moon, 198, 225
Newton, Isaac
and gravity, 139, 175, 231
and the moon, 6, 304
Nova, 6, 28, 174, 205–207, 214, 253, 260

O
Olber's Paradox, 253, 256
Oort Cloud, 139, 145

P
Parallax, stellar, 111, 118, 175
Pepys, Samuel (1633–1703), 148
Phobos–Grunt, Russian mission, 200
Piccolomini, Alessandro (1508–1579), 301
Pickering, Edward Charles (1846–1919), 30
Pleiades. See M45
Pluto, 2, 53, 101, 122, 192, 193, 196, 197,
    289, 310
Podgson, Norman (1829–1891), 31
Pritchard, Charles (1808–1893), 28
Ptolemy, Claudius, 11, 12, 30, 77

R
Rasselas, 236
Regiomontanus (Johannes Muller von
    Konigsberg, 1436–1476), 76
Ringwood, Stephen, 42
Roeslin, Helisaeus (1544–1616), 214
Römer, Ole (1644–1710), 111, 174
Rosse, Lord (William Parsons, 1800–1867), 50

S
Sacrobosco, John of, 11, 43, 302
Saturn
    moons, 109, 241–243
    observing, 239–243
    rings and divisions, 197, 239–242, 246
Saxony, Albert of, 11
Scheiner, Chrisoph (1573–1650), 20
Schiaparelli, Giovanni (1835–1910), 273
Schwarzschild radius, 232, 234
SDO, NASA mission, 25, 144
Secchi, Angelo (1818–1878), 30
Seeing Scales, 29
SETI, 96, 129, 137, 138, 222
Seyfert, Carl Keenan (1911–1960), 256
Shapley, Howard (1885–1972), 71
Skylab, NASA mission, 282

Slipher, Vesto (1875–1969), 256
Snow, C.P. (1905–1980), 250
SoHo, NASA mission, 25, 144
Spectroscopy, 15, 17, 54, 60, 71, 156, 200,
    289–290
Sputnik, 278
Stardust, NASA mission, 147
Stradivarius, 26
Struve, Otto (1897–1963), 260
Struve, Wilhelm (1793–1864), 110
Sun
    chromosphere, 21, 22
    corona, 15, 18, 22–24, 144, 183, 207, 282
    photosphere, 19–23, 105, 236
    safe observing, 17, 19, 20
    solar wind, 10, 15, 23–26, 63, 126, 225,
        304, 307
Sunspots, 15, 19–21, 25, 95, 290, 308
Supernova, 59, 71, 95–97, 136, 205–215, 260,
    263, 264
Surveyor 7, NASA mission, 10, 126

T
Tardigrades, 284, 285
Telescope, invention, 4, 5, 41, 101, 161
Themis, NASA mission, 26
Tisserand, Francois Felix (1845–1896), 144
Tissint meteorite, 82
Titan, moon of Saturn, 242, 248
Tombaugh, Clyde (1906–1997), 192
Trans-Neptunian objects, 192
Transpermia, 200
Tsiolkovsky, Constantin (1957–1935), 278

U
Uhuru, NASA mission, 231
Ulysses, NASA mission, 23
Uranus
    discovery, 217
    internal structure, 221, 248
    moons, 223
    observing, 219–220
    rings, 218, 223

V
Variable stars
    cataclysmic, 205, 207
    eruptive, 260
    labelling conventions, 60, 131, 154, 181,
        205, 211, 223, 253, 300
    types, 104, 153, 154, 159, 165

Venera, Soviet missions, 270
Venus
    Galileo and phases, 20, 42, 43, 267
    observation, 265, 266
    transits, 269
Viking NASA missions, 84, 85
Voltaire (1694–1778), 113
Voyager 1 and 2, NASA missions, 46, 63

W
Wells, HG (1866–1946), 82
White Dwarf stars, 103, 108, 109, 131, 209
WMAP, NASA probe, 136, 261

Z
Zwicky, Fritz (1898–1974), 260